Seven Saints of Science

Seven Saints of Science

An Introduction to the History of Conflict
Between Science and Religion

Richard C. Powell
Doug Powell

Jenny Stanford
Publishing

Published by

Jenny Stanford Publishing Pte. Ltd.
101 Thomson Road
#06-01, United Square
Singapore 307591

Email: editorial@jennystanford.com
Web: www.jennystanford.com

British Library Cataloguing-in-Publication Data
A catalogue record for this book is available from the British Library.

Seven Saints of Science: An Introduction to the History of Conflict between Science and Religion

ISBN 978-981-5352-24-5 (Hardcover)
ISBN 978-1-042-00201-6 (eBook)
DOI: 10.1201/9781042002016

Contents

Preface

This book provides a unique look at the relationship between science and religion from the dual perspectives of one author with a PhD in physics and 40 years of experience in research and teaching and another author who has an MA degree in apologetics (PhD in progress) with extensive writing on religion. The book discusses how the fields of science and religion interact and argues that they are not antithetical to each other as might be concluded from stories in popular literature. It shows how the timing of the development of the two disciplines and new ways of thinking led to conflict between science and religion. That development is traced in the changes in society as it moved from the Renaissance through the Age of Enlightenment and the Industrial Revolution, the Victorian era, the age of the new physics, World War II, and into the high-tech era. Biographies of leading scientists from each of these eras who also had strong religious faith are used to demonstrate the compatibility of science and religion. Each of these scientists was quite different from the others, but their faith was a fundamental attribute that grounded their thinking and matched the time in which they lived. They are not outliers or fringe thinkers or misguided zealots who tried to use the Bible as a scientific text. Rather, they are luminaries who significantly advanced human knowledge believing that their contributions not only illuminated the world but also had a divine source. Where modern society often sees conflict and competition, these visionaries saw harmony and compatibility. What was it these men saw that seems to elude so many today? That is what this book tries to answer.

Chapter 1

Introduction

1.1 Science and Religion

For since the creation of the world his invisible attributes—his eternal power and divine nature—have been clearly seen, because they are understood through what has been made. So people are without excuse.

Romans 1:20

Scripture and creation, the two books of revelation

Seven Saints of Science: An Introduction to the History of Conflict between Science and Religion
Richard C. Powell and Doug Powell

ISBN 978-981-5352-24-5 (Hardcover), 978-1-042-00201-6 (eBook)
www.jennystanford.com

DOI: 10.1201/9781042002016-1

Christianity is a revelatory religion. God reveals himself to mankind in two ways: scripture (special revelation), and creation (general revelation). Metaphorically, this has been described as two books written by the same author. When each is properly understood, they are compatible with each other, in perfect harmony. And when they are in conflict, then one or both of them are misunderstood. However, there is a tendency to separate them into two different fields of knowledge: science and religion.

Science and religion are two important aspects of society and culture. Both are gifts of God based on humankind being given the ability to reason. Over the years, historians, philosophers, and theologians have pointed out that they interacted with each other in four different ways: conflict, independence, dialogue, and integration.[1,2] The Conflict view sees science and religion as inherently unreconcilable where only one authority can exist. The Independence view compartmentalizes the two kinds of knowledge so that they cannot be mingled; good fences make good neighbors. The Dialogue view sees science and religion as different frequencies of light, each illuminating the world in different ways to increase the vision of both. And the Integration view sees them as inextricably linked, where religion is the blueprint for what exists in the world.

In the early 2000s, a number of vocal atheists such as biologist Richard Dawkins, neuroscientist Sam Harris, philosopher and cognitive scientist Daniel Dennett, and writer Christopher Hitchens popularized the *conflict* view, emphatically reinforcing the misperception it was the primary view of scientists. The National Academy of Science has adopted the point of view of *independence,* that science and religion should be separate so there is no conflict.[3] However, religious leaders like Pope John Paul II believe *dialogue* between science and religion is important because "science can purify religion from error and superstition; religion can purify science from idolatry and false absolutes."[4] As we will find in the following chapters, many great scientists believe science and religion should be in dialogue or are integrated with each other.

Richard Dawkins (Photo by Darwins Rottweiler CC-BY-3.0).

Even though *science* and *religion* are fundamental terms for navigating our current times, we often assume everyone agrees on what they mean. But both words can be hard to define, so it's important to be clear on what we mean before we go any further.

Science is the study of nature in the world around us. It includes both the structure and behavior of everything in the universe God created. Knowledge is considered to be scientific if it was gained through evidence obtained by a set of specific techniques including observation, experimentation and testing. The scientific method for obtaining knowledge generally has seven steps:

1. Question
2. Research
3. Hypothesis
4. Experiment
5. Data Analysis
6. Conclusion
7. Communication

The final step is important in assuring that the knowledge found by a scientist is accepted by their colleagues and the general public. The scientific method can be applied to almost any important problem for fact-based decision making.

After the explanation of something is well established by the scientific method, it can be called a scientific theory. A theory is an explanation of something that uses facts to verify a hypothesis. It is testable. It should be supported by the results of every new test. If the data from new tests do not support the theory, a new theory should be developed.

Religion is a system of belief in God or gods. This usually includes implications for morality and ethics. Sometimes the term *religion* is used as a subset of anthropology to study what different people believe about God. This is totally different from *theology*, which is the study of God himself (not what people believe about him). In this book, we are using *religion* and *theology* interchangeably, although the scientists we look at don't all agree on who God is. And because the scientific revolution occurred in the context of Christendom, the specific religion in view most of the time is Christianity.

Michelangelo's creation of Adam.

One of the reasons underlying the perceived differences between science and religion is that they developed at different times in history. By the time Nicholas Copernicus published *On the Revolutions of the Heavenly Spheres* in 1543 launching the scientific revolution, essential Christian theology had been established for more than a thousand years.[5] The church split in 1054, separating the Eastern tradition (Orthodoxy) from the Western tradition (Roman Catholicism). Already, the religious power of the Catholic church had become increasingly intertwined with the political structure of Europe. Copernicus announced his findings at the same time Martin Luther and John Calvin were publishing the works that became the founding documents of the Protestant Reformation, and at the same time the Renaissance saw Leonardo da Vinci and Michelangelo at the full heights of their creative genius.

God the Geometer from Codex Vindobonensis (c.1220–1230).

Clearly, the theology and traditions of the church developed in pre-scientific times. The development of science was inhibited by the lack of the number zero in the Western numerical system. Eastern cultures had not only developed a number *zero* but also preserved much of the philosophical and scientific thought developed by the ancient Greeks such as Aristotle. Although the early church fathers were heavily influenced by classic Greek philosophers as they developed their theology, eventually the writings of the philosophers were mostly lost to the West. During the reign of Justinian (527–565), the schools of philosophy in Athens relocated to the East. With the advent of Islam, Aristotle and other important Greek thinkers became preserved largely by Islamic philosophers. The Muslims brought these works to the Iberian Peninsula beginning with their invasion in 711. During the predominantly Roman Catholic reconquest that drove the Muslims out of Spain by 1492, they left behind libraries with works on mathematics and copies of ancient Greek philosophy. After Toledo was retaken in 1085, the archbishop established a translation center at the cathedral to introduce Latin versions to the West, leading to the rediscovery of the Greek classics, including—most importantly—Aristotle. These discoveries provided essential concepts that helped drive the new scientific age.

The new and newly recovered information sparked the Renaissance in the 15th and 16th centuries, a cultural transition out of the Middle Ages. The Renaissance was focused on art, architecture, music, and philosophy, but also stimulated free-thinking that later impacted science. The Renaissance matured into the Age of Enlightenment (c.1685–c.1815), a period marked by the birth of modern science and its impact on society. The Age of Enlightenment, in turn, spawned the Industrial Revolution (1750–1850). The scientific discoveries and inventions that occurred during this time not only improved society but challenged it with many new ideas beyond the strictly scientific. The scientific community became an important contributor to society. Although many scientists believed their work revealed the mind of God, the increase of knowledge about the world also fostered the idea that the world could be understood apart from God. England was one of the centers of scientific development, and the scientific community became a powerful influence there during the Victorian Era (1837–1901).

By the end of the 19th century, most scientists believed that all the basic concepts of science were known, and the future would just involve making more accurate measurements. However, during the first quarter of the 20th century the birth of modern physics occurred, changing the paradigm. Scientists gained knowledge of sub-atomic particles, motion near the speed of light, and the origin of gravity. Along with this new knowledge came new basic concepts of measurement uncertainty, wave-particle duality of matter, and discrete rather than continuous states of existence. These new ways of thinking influenced all aspects of society including philosophy and religion. This was followed by the period from 1939 to 1945 which was consumed by World War II. Science played a key part during the war with developments such as radar and nuclear energy. The period since the end of World War II is considered the modern scientific era. Scientific discoveries have had a major impact on society during this period, including microelectronics based on the transistor, optical communication, medical procedures based on lasers, television (which grew out of radar technology), and clean energy sources such as solar cells and nuclear fusion.

One of the important philosophical developments of the scientific era was empiricism, which elevated scientific knowledge above other kinds of knowledge. Empiricism claims that the only way to obtain knowledge is through the senses and experience. Observations, collecting data, and interpreting it led to a process of data-driven decision making. Rationalism (the belief that knowledge can come from sources other than sense experience) fell out of favor. And the concept of revealed religion—that God had spoken—was dismissed as subjective, irrelevant, or the realm of intuition. This led to problems in considering anything that was outside the scope of direct observation and experience. Miracles, for example, cannot be explained scientifically (although the effects of some miracles may be testable in theory). And ironically, most of creation consists of dark energy and dark matter that cannot be observed. In these cases, interpretation based on reason must be used. And since interpretation of any data involves reason, reason can't be rejected completely as a legitimate means of knowledge.

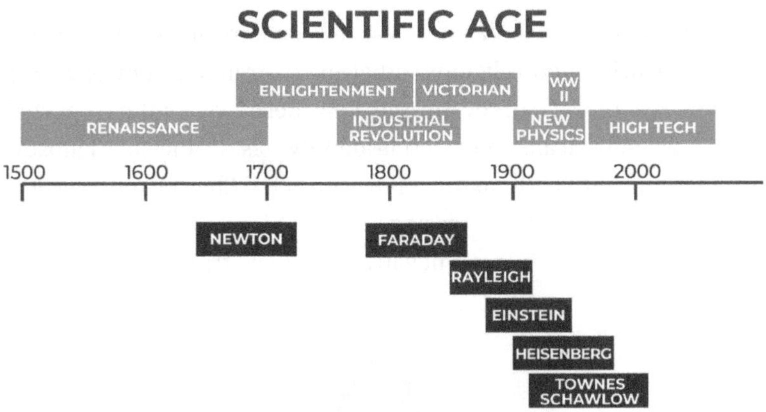

Scientific age timeline (© Doug Powell)

Another problem with empiricism was that data might be interpreted in several different ways by different people, or new experiments would produce different data. Thus, empiricism could not be the solution to everything.

1.2 The Rise of Atheism

In 399 BC, Greek philosopher Socrates was executed for rejecting the Greek pantheon of gods and was therefore seen as corrupting his students, which included Plato, who in turn was Aristotle's teacher. For Socrates, atheism—the rejection of the existence of God or gods—was literally deadly business. Although atheism has existed throughout human history, it has not always been acceptable to profess. When it has been expressed, it has taken different forms. Agnosticism holds that even if God does exist, we can't know anything about him, or that there is a lack of evidence. Deism allows God to create the world and wind it up like a clock but then does not engage with it. Philosophy—particularly since the Age of Enlightenment—is often a document of well-articulated skepticism. Although certain religions such as Buddhism, Shintoism, Taoism, Confucianism, and arguably Hinduism are atheistic and can skew the measurable percentage of atheists in the world, non-religious atheism is trending upward. For example, a recent survey by the Pew Research Center reported 28% of Americans now self-identify as "atheist, agnostic, or nothing in particular."[6] But as G.K. Chesterton warned, "Atheism...is the most daring of all dogmas...for it is the assertion of a universal negative."[7] What triggered atheism's growth?

Atheist symbol

During the Middle Ages, cathedral schools gave birth to the first universities. The different academic disciplines were united by the belief that theology was the science that united all other thought and eventually came to be called the Queen of the Sciences. Thomas Aquinas referred to theology as the "nobler"[8] science because it considered the "first and universal causes" and the "universal end of all things."[9]

Aquinas was far from unique in his thinking. Long before his time, ancient cultures did not distinguish between religion and primitive science. Ancient priests not only officiated worship; they also calculated *when* to worship, which meant reading the stars. This carried over to divining when to undertake any important endeavor such as planting crops, holding a wedding, and going into battle.

Nicolaus Copernicus (1473–1543)

Galileo Galilei (1564–1642)

The first major conflict between the church and science was foreseen by Copernicus, a devout Catholic who—aside from being an astronomer—served as a canon at the cathedral in Frombork, Poland. Although he published a short work on heliocentrism (the Earth moving around the Sun) in 1514, he resisted releasing his ground-breaking full treatment on the subject, *On the Revolutions of the Heavenly Spheres*, until shortly before his death in 1543, fearing it would cause controversy. In the dedication he included to Pope Paul III, he wrote, "Perhaps there will be babblers who claim to be judges of astronomy although completely ignorant of the subject and, badly distorting some passage of Scripture to their purpose, will dare to find fault with my undertaking and censure it."

The prevailing view Copernicus challenged came from Aristotle, who held the stars and planets revolved around the Earth as an unchanging universe, and Ptolemy (c. AD 100–170), who further developed Aristotle's formulations. Although Copernicus did not run afoul of the church, his advocate Galileo Galilei (1564–1642) did. He, too, was a committed Christian who believed, "God is known...by Nature in His works, and by doctrine in His revealed word."[10] After the invention of the telescope in 1608, Galileo built his own to study the night sky. He published his observations in 1610 in a book called the *Starry Messenger.* In it, he details the phases of the moon and planets, which show they are in motion around the Sun, providing evidence for Copernicus' heliocentric view. Because the Aristotelian/Ptolemaic view had become so ingrained in the church, to challenge geocentrism was to challenge the church. Galileo recognized that the church had not only mingled scripture with science, but that it was bad science. He wrote, "The intention of the Holy Ghost is to teach us how one goes to heaven. not how heaven goes."[11] In 1616, the church forbade him from teaching heliocentrism. The fears of Copernicus came true in the very church he gave the warning to.

Galileo published the more philosophical *Dialogue Concerning the Two Chief World Systems* in 1632. The following year, he was tried for suspicion of heresy, convicted, and placed under house arrest for again teaching Copernicus' theory that the Earth revolved around the Sun.[12]

Rather than serving as a cautionary tale, this controversy became emblematic of the difficulties arising from how to affirm and reconcile scientific and religious knowledge. But the biggest factor in segregating the two ironically came from an attempt to preserve them.

Immanuel Kant (1724–1804).

The work of Immanuel Kant (1724–1804) has been called the Copernican Revolution in philosophy. He agreed with the empiricists who believed knowledge was acquired through the senses. However, he understood that if knowledge was limited to the senses, it would result in chaos because morality—laws that govern behavior—don't come from sense experience. Therefore, morality must be a different kind of knowledge grounded in a non-physical, eternal realm of ultimate reality. Kant called the physical realm we encounter with our senses the phenomenal world. Since we only know it through our senses, we can only know how the world appears to us, not how it actually is. He called the immaterial realm the noumenal world, the place of ultimate reality which is known by intuition and can't be known through our senses (e.g., morality, logic, categories). The noumenal world includes God because the moral law requires a transcendent moral law-giver to ground it. Although Kant rejected rational arguments for God because they reason from sense experience, his system, nevertheless, works as an argument from morality for God. Kant also distinguished the two realms by defining *knowledge* as applying to the phenomenal world, and *belief* as applying to the noumenal world. He wrote, "I must...abolish knowledge to make room for belief."[13] Thus, Kant created a structure where scientific knowledge became qualitatively different from religious belief.

Despite the split, there was still a problem for scientists who had no religious conviction: Science did not explain the origin of life. God was still conceptually necessary as a precondition for the universe. As scientific knowledge increased, its explanatory power transformed physics, chemistry, astronomy, and all other areas of natural science with one exception: biology. At the beginning of the 19th century, many scientists believed living beings contained a complexity and precision that could only be the result of design, which implies a designer. However, a growing number of scientists rejected that view for being unscientific and looked for a naturalistic explanation for how life came to be.

Enter Charles Darwin (1809–1882). In 1859, he published *On the Origin of Species*, which theorized the mechanism of natural selection as an explanation for how species evolve. Natural selection says variations occur in offspring, some giving an advantage over others allowing them to best adapt to their environment. The advantaged offspring survive and pass on the advantage to their offspring with their own variations so that the accrual of advantages ensures the survival of the fittest.

Charles Darwin in 1880

For atheists such as Thomas Huxley, Darwin's staunchest early defender, evolution through natural selection was the final piece of the puzzle for naturalism. God was no longer necessary as an explanation. At this time, confessing to atheism carried a stigma, which was why Huxley coined the term *agnostic,* meaning that even if God did exist, nothing could be known about him, and was therefore irrelevant. In the wake of Darwin, atheism began to be accepted as respectable.

As a result, scientific naturalism—the belief that science has or will have the ability to explain all phenomena in the natural world—developed into its own philosophy (which ironically cannot be verified by science since it is a product of reasoning). Those who hold to scientific naturalism generally believe science is the highest—if not only—form of knowledge, relegating all other claims of knowledge to some measure of inferior knowledge or at least subjective opinion. This is one side of the battle line in the contemporary conflicts between science and religion.

(© Doug Powell)

For those who have religious belief, there is a place for *methodological* naturalism, which is the presupposition to not appeal to God as a scientific explanation. At the same time, methodological naturalism allows for scientific evidence to infer God's existence (or at least an intelligent designer) if and when the appearance of design is detected. Whether or not evidence of design is truly present is a factual dispute open to debate without dismissing the implications of a designer's existence out-of-hand.

A wide range of issues such as evolution, the Big Bang, genetic research, sexuality, and the place of religious expression in the public sphere have all become battles where science and religion are too often pitted against each other. And too often, the enemy targets are strawmen attacked by poor representatives of their causes. Lawsuits and the politicization of certain topics muddy the waters even further leading to the common misperception that science and religion are incompatible.

So what is a better way forward? Without offering a miracle cure that resolves the issue, there are certain principles that would cool the rhetoric and return it to a place where ideas can be exchanged in a way where each realm speaks most clearly, shedding light into areas where the other realm cannot reach.

For Christians who hold the Bible as inerrant, this means approaching scripture keeping in mind the intention of the author and who the original audience was. The Bible was written in a pre-scientific age to pre-scientific people. Science was not a category of knowledge for them. When the Bible makes a statement that falls within the realm of science, then if the Bible and science are properly understood, there will be no conflict. But when the Bible is treated as a scientific explanation for the world, interpretation becomes eisegesis (reading meaning *into* the text rather than finding the meaning *of* the text). An important consideration for the text is recognizing different genres, metaphors, poetry, and parables. An understanding of surrounding cultures is helpful background information for recognizing polemical motivations for some passages. Misinterpretation—including a literalistic interpretation—is a contributor to some of the conflicts.

Christians also need to keep in mind that the Bible is not an exhaustive compendium of truth. As the Westminster Shorter

Catechism states, "The scriptures principally teach what man is to believe concerning God, and what duty God requires of man."[14] Christians sometimes confuse having the truth of God with having an exclusive claim on truth, goodness, and beauty. But the doctrine of common grace expresses the reality that all people—even unbelievers—can discover great truths, live virtuous lives performing acts of great goodness, and create profoundly beautiful works of art. That includes recognizing and accepting scientific truth regardless of who discovers it.

Theologians throughout church history have understood this. Many early church fathers believed pre-Christian thinkers such as Socrates, Plato, Aristotle, Homer, Plutarch, Thales, and Pythagoras had such great insight in the truths of reality that they were considered believers in the faith not yet fully revealed. That's why Dante, in his *Divine Comedy*, places them in the outermost circle of Hell, which is a place of limbo and suffering without torment.[15] And it's why icons of these Greek thinkers are often included in Orthodox monasteries, depicted surrounding Jesus. They have even become known as "Virtuous Pagans."

Some Christians have also misunderstood the meaning of biblical faith. Faith is not a synonym for hope, wish, or desire. And faith is not the opposite of reason and is not incompatible with evidence. The Greek word in the New Testament that is translated into English as *faith* is the word *pistis*, which means firm persuasion and conviction.[16] What do you need to be persuaded? Reasons. What do you need to be convicted? Evidence. Admittedly, not all reasons and evidence are within the purview of science, but it certainly has an important place that is entirely compatible with the New Testament's concept of faith.

For scientists, the key challenge is acknowledging that theirs is not the only legitimate form of knowledge, and that science is incapable of full explanatory power for the world and our experience in it. To ignore this truth is to create a kind of modern priesthood where scientists are the sole arbiters of truth. Yet the natural sciences leave no room for speaking to issues of morality and ethics. They cannot account for universals (such as numbers and categories), the laws of logic, or even the natural laws themselves. When scientists speak outside the confines of science, conflict with religion often arises.

With this in mind, we are going to discuss the lives of seven great scientists who held deep religious convictions: Isaac Newton, Michael Faraday, Lord Rayleigh, Albert Einstein, Werner Heisenberg, Charles Townes, and Arthur Schawlow. Although they were never canonized as saints by either the Catholic or Orthodox churches, all of them but Einstein were pious, observant Christians (Newton's heterodoxy notwithstanding). Einstein constructed his own faith which coupled with his scientific insights in a tradition similar to the Virtuous Pagans. If there were such a thing as sainthood among the scientists, surely these seven men would be among them. They cover the time in history during which the antagonism between science and religion developed.

Despite all being men of religious faith, each of these scientists was quite different from the others. Their methods of scientific inquiry were based on experimental demonstrations or mathematical formalism, or creative thinking that matched the era of history in which they worked. They came from different parts of society: peers of the realm; commoners; and the poor. They had very different religious views, ranging from traditional denominations to small sects with niche beliefs to Einstein's invention of his own religion, but all of them held to a faith that complemented science. Even with their different perspectives on religion, they shared the belief that doing scientific research is learning something about God. The most important thing to know about these people is that each of them viewed science through the lens of religion. We experience this every Sunday at my (Richard's) church, which displays the cross on our altar before a great window with God's creation in the background.

Not only did science undergo significant changes in these historic eras, but the church also changed significantly. Each chapter tells the story of how a great scientist combined their scientific work with their religious beliefs, describes how the church changed during this period, and what happened with conflicts between science and religion.

> *"Do not be afraid of being free-thinkers. If you think strongly enough you will be forced by science to the belief in God, which is the foundation of all Religion. You will find science is not antagonistic but helpful to Religion."*
>
> Lord Kelvin (1824–1904)
> Letter to *The Times,* May 2, 1903[17]

St. Philips in the Hills Episcopal Church, Tucson, Arizona (© Richard Powell)

Raphael *School of Athens*

References and Notes

1. Barbour, Ian G., *When Science Meets Religion: Enemies, Strangers, or Partners*? (New York: HarperCollins Publishers, 2000), 703.
2. Ruse, M., McGrath, Alister E., and Gordon, B. L., *Three Views on Christianity and Science* (Grand Rapids: Zondervan Academic, 2021), 14.
3. *Science, Evolution, and Creationism* (Washington D.C., National Academies Press). DOI: https://doi.org/10.17226/11876.
4. Letter of His Holiness John Paul II to Reverend George V. Coyne, S.J., Director of the Vatican Observatory, 6.1.1988, https://www.vatican.va/content/john-paul-ii/en/letters/1988/documents/hf_jp-ii_let_19880601_padre-coyne.html (accessed 11.30.24).
5. The Fourth Ecumenical Council, the Council of Chalcedon, was held in AD 451. With the three councils that preceded it, it established and affirmed the Trinity and the dual nature of Christ (hypostatic union). These doctrines are expressed in the Nicene Creed and the Definition of Chalcedon.
6. Has the rise of religious 'nones' come to an end in the U.S.? (https://www.pewresearch.org/short-reads/2024/01/24/has-the-rise-of-religious-nones-come-to-an-end-in-the-us) (accessed 12.02.24).
7. Chesterton, G.K., *Twelve Types* (London: Arthur L. Humphries, 1906), 95.
8. Aquinas, T., *Summa Theologica* I.1.5.
9. Aquinas, T., *Commentaries on Aristotle*, Book 1, Lesson 2.
10. Galileo, Letter to the Grand Duchess Christina of Tuscany, 1615. (https://history.hanover.edu/courses/excerpts/111gal2.html) (accessed 11.30.24).
11. Ibid.
12. De Santillana, G., *The Crime of Galileo* (Chicago: University of Chicago Press, 1955), 310. De Santillana provides an English translation of the Inquisition's condemnation.
13. Kant, I., *Critique of Pure Reason*, 1781.
14. Westminster Shorter Catechism, Answer to Question 3: What do the scriptures principally teach?

15. Dante, *Divine Comedy, Inferno,* Canto IV.
16. *Mounce Concise Greek-English Dictionary of theNew Testament,* Mounce, W., and Bennet, R.D. eds. (Oak Tree Software: 2011).
17. Thompson, S.P., *The Life of William Thomson, Baron of Kelvin of Largs* (London: Macmillan, 1910), 1099.

Chapter 2

Isaac Newton (1642–1726): Under the Apple Tree

All things in heaven and on earth were created in him—all things, whether visible or invisible,...all things were created through him and for him. He himself is before all things and all things are held together in him.

Colossians 1:16–17

"And from his true dominion it follows that the true God is a living, intelligent, and powerful Being; and, from his other perfections, that he is supreme, or most perfect. He is eternal and infinite, omnipotent and omniscient; that is, his duration reaches from eternity to eternity; his presence from infinity to infinity; he governs all things, and knows all things that are or can be done."[1]

Isaac Newton

Seven Saints of Science: An Introduction to the History of Conflict between Science and Religion
Richard C. Powell and Doug Powell

ISBN 978-981-5352-24-5 (Hardcover), 978-1-042-00201-6 (eBook)
www.jennystanford.com

DOI: 10.1201/9781042002016-2

Isaac Newton (© Doug Powell).

2.1 Newton's Science and Faith

Have you ever sat under an apple tree in September and watched the apples fall? Over 400 years ago, Isaac Newton did, and he asked himself, "Why do the apples always fall down and never up?" Today this may seem like a silly question, but in the 1600s, people had no idea what gravity was or how it worked. The word *gravity* hadn't even been coined. So it wasn't a dumb question, but only Isaac Newton had the curiosity to try to answer it. Newton's curiosity led him to develop a mathematical theory describing how two bodies, each with some mass, exert a force of attraction on each other, acting over the distance between them. This classical theory of gravity is still useful in explaining what we observe in normal, everyday life associated with the effects of gravitational attraction. Even though more sophisticated views of gravity were developed by Einstein to explain detailed observations outside the realm of classical physics, the Newtonian theory of gravity remains the best way to understand most of our observations.

Isaac Newton in 1689 by Godfrey Kneller

Newton's Tree at Woolsthorpe Manor (Photo by Bs0u10e01 CC-BY-SA4)

A fire in Newton's laboratory

What made Isaac Newton curious enough to think about the way gravity works? To achieve success and impact society, a scientist's work must be compatible with the time during which they live. In this case, Newton lived during one of the most important eras in the history of our civilization, called the Age of Enlightenment (c.1685–1815). The Renaissance brought the art and literature of Europe out of the Middle Ages through an intense period of renewed creativity in the 15th and 16th centuries inspired by the recovery of classical Greek culture and philosophy. Near the end of the Renaissance, a scientific revolution occurred, led by luminaries such as Copernicus and Galileo. Newton was born the year Galileo died and was destined to carry on much of the work Galileo started. This ascendence of science and philosophy characterized the Age of Enlightenment. During this period, people began to pursue knowledge following scientific inquiry and reason rather than accepting on authority what they were told by others. Immanuel Kant, one of the most important thinkers of the time, declared the motto of the Enlightenment was, "Have courage to use your own reason!"[2] As a result, new ideas began to develop. Individual liberty became more important than allegiance to a monarch, while Protestant doctrines were derived from the study of scripture rather than received from the traditions and decrees of the Roman Catholic church. This was the time of the American Revolution, the French Revolution, and the attempted English Revolution, marking the rise of democracy. Much of what we value in society today had its beginnings in the Age of Enlightenment.

One of the major changes during the Enlightenment was the establishment of scientific societies. Before this time, very few people were able to learn about scientific discoveries and ideas. Even though the printing press was invented in the 1440s, mass printing of information was not common for another couple of hundred years. The Royal Society of London was established in 1660 and published its first journal four years later. Other European countries established scientific societies or academies around this time. The main purpose of these organizations was to broadly disseminate scientific information. The communication of new knowledge and discoveries to scientific colleagues and the general public was a critical development for the scientific era.

This was the perfect time in history for a person with an active curiosity. And the leading scientist of this era was Sir Isaac Newton, a man whose unconventional thinking matched the attitude of society in this era. Along with developing the theory of gravity, Newton made major contributions in classical mechanics, optics, astronomy, mathematics, and alchemy, and also wrote voluminously on theology.

Isaac Newton was born in Woolsthorpe-by-Colsterworth, England on Christmas Day of 1642 into a Puritan family. (Puritans were English Protestants who applied the theology of the Reformation to the Church of England.) His father died before he was born, and his mother remarried and moved away, leaving him to be raised by his maternal grandmother. He hated his stepfather and resented being deserted by his mother. This traumatic experience as a young child had a significant effect on his personality. He grew into an angry, contentious person who was easily slighted—a difficult temperament that plagued him throughout his life. Newton was educated at The King's School in Grantham and graduated as the top-ranked student. However, his education was paused when his stepfather died and his mother moved back home. She insisted he manage the work on her farm instead of going to school. He hated this and failed miserably at his work, so his mother relented and sent him back to school.

Newton demonstrating an optical experiment

It is not clear how much Newton's early life was affected by the English Civil War that was going on at this time. After beginning in 1642 and the beheading of King Charles in 1649, Oliver Cromwell and his son Richard ruled as Lord Protectors until 1659. King Charles II was enthroned in 1660, restoring the monarchy. Newton himself was not directly involved with these events, but the civil war freed institutions such as the University of Cambridge from the control of the church.

After he finished high school, Newton attended Trinity College at the University of Cambridge and obtained his B.A. degree in 1665, just as the Great Plague hit Europe. Everything closed down, including the university, and for the next two years, Newton continued his education at home in Woolsthorpe. During this time, he made great progress working by himself on several diverse topics such as gravity, optics, and calculus. It turned out to be a highly productive period for him despite domestic interruptions such as his dog tipping over a candle and burning his papers.

In 1667, society reopened and Newton returned to Cambridge and was elected a Fellow of Trinity College. However, at this time Fellows of Cambridge were required to be ordained priests or pledge their adherence to the Church of England. Newton refused to do this because his own religious beliefs were not consistent with traditional Anglicanism. However, he was granted an exception by King Charles II so he could accept the position. Newton received his M.A. degree, became the Lucasian Professor of Mathematics at the University of Cambridge, and was elected a Fellow of the Royal Society (which is like the U.S. National Academy of Sciences). He never married.

During his life, Newton made contributions in three very different areas: science, public service, and theology. As a scientist, Isaac Newton's achievements spanned broad areas of math and physics. For example, he was one of the major developers of the mathematical field of calculus which is important in all areas of science and engineering. Much of his mathematics is documented in his famous book *Philosophiæ Naturalis Principia Mathematica* (nicknamed *Principia*). Published in 1687, it is considered to be one of the most important works of modern science. He explained the reason for writing the book in a letter from 1692. "When I wrote my treatise about our system, I had an eye upon such principles as might work with considering men, for the belief of a deity, and nothing can rejoice me more than to find it useful for that purpose."[3] In other words, Newton strongly believed in God who created the universe we live in.

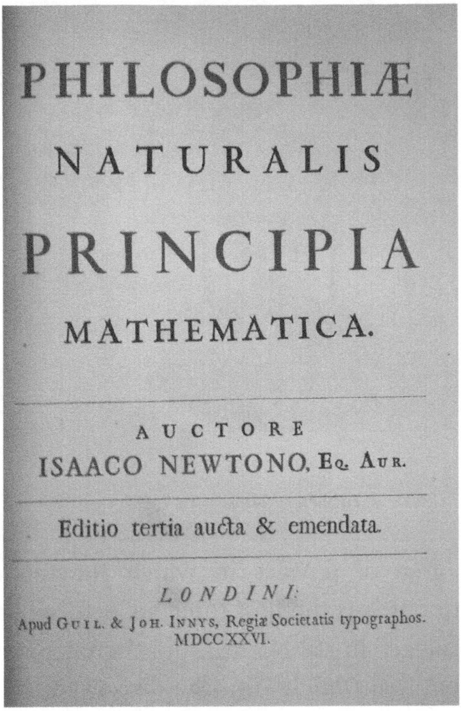

PHILOSOPHIÆ
NATURALIS
PRINCIPIA
MATHEMATICA.

AUCTORE
ISAACO NEWTONO, Eq. Aur.

Editio tertia aucta & emendata.

LONDINI:
Apud Guil. & Joh. Innys, Regiæ Societatis typographos.
MDCCXXVI.

Title page of Newton's *Principia*

It was in *Principia* that he described the physics of motion covering what we now call *classical mechanics*, based on Newton's three laws of motion. These three laws describe how a force changes the motion of an object. The first law states that the motion of an object will not change unless a force is applied to it. The second law says that the acceleration of an object depends on the force applied to it and its mass. The third law says that for every force there is an equal and opposite force. We still use these three laws of physics, and the unit of force is called a Newton.

A replica of Newton's telescope

Another area of physics in which Newton had a great impact is optics. It was the first topic he lectured on as a Fellow of Trinity College. In 1704, he published a classic textbook in this area entitled *Opticks*. In it he discussed three important aspects of optics that are still in use today: the theory of color, telescopes, and beam expanders. Newton studied how white light going through a prism is separated into different colors. Using this observation, he defined the scientific nature of color and the composition of light. He also demonstrated how light could be separated into colors using reflection as well as refraction. Using a lens and a flat piece of glass separated by a thin film of air, he produced what are still known as Newton's rings. In another experiment, he invented the reflector telescope. Up until that time refractor telescopes had been made using lenses. He demonstrated that using mirrors instead of lenses produced clearer images. His designs form the basis of modern telescopes used in astronomy today. Also, in *Opticks* he described the use of multiple prisms to expand a beam of light. Today his concept of a beam expander is used in the design of high-power lasers.

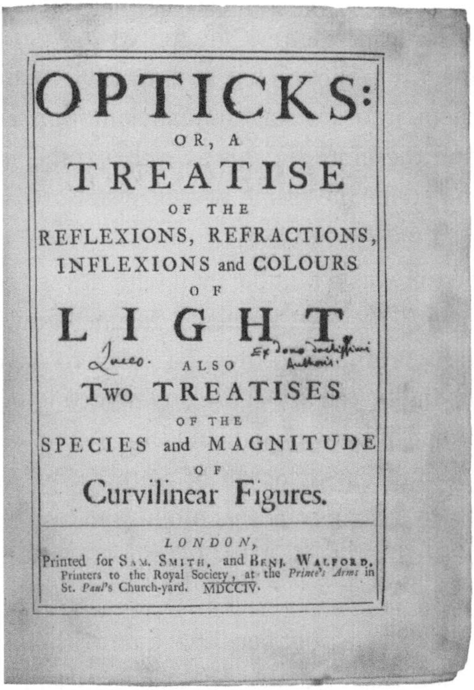

OPTICKS:
OR, A
TREATISE
OF THE
REFLEXIONS, REFRACTIONS,
INFLEXIONS and COLOURS
OF
LIGHT.
ALSO
Two TREATISES
OF THE
SPECIES and MAGNITUDE
OF
Curvilinear Figures.

LONDON,
Printed for SAM. SMITH, and BENJ. WALFORD, Printers to the Royal Society, at the *Prince's Arms* in St. *Paul's* Church-yard. MDCCIV.

Title page of Newton's *Opticks*

Newton's theory of the physics of gravity used mathematics he developed to describe the effects of the earth's gravity on objects on the earth's surface as well the describing the motion of the moon and planets in the solar system. He was the first person to use the word *gravity* to describe the interaction between two objects. Newton also spent much of his time with chemistry in the field of alchemy, but he left no major achievements in this area.

During his lifetime, some scientists used Newton's laws of motion and theory of gravity to describe the universe as a machine. Because of his belief in God as the creator of the universe, Newton did not agree with this characterization of his work. In a series of famous letters between Newton and Dr. Richard Bentley he helped Bentley develop his sermons at St. Martins in the Field church using *Principia* to argue against atheism. In one letter, Newton wrote, "So then gravity may put planets into motion but without divine power it could never put them into such a Circulating motion as they have about Sun, and therefore for this as well as other reasons I am compelled to ascribe the frame of this system to an intelligent agent."[4]

Isaac Newton was a humble man who never bragged about his achievements. He once wrote, "I do not know what I may appear to the world, but to myself I seem to have been only like a boy playing on the sea-shore and diverting myself in now and then finding a smoother pebble or a prettier shell than ordinary, whilst the great ocean of truth lay all undiscovered before me."[5] He also wrote the often-used quote, "If I have seen further [than others], it is by standing on the shoulders of giants."[6]

Newton was also a very sensitive man with an unattractive personality that was not able to take criticism of his work graciously. The scientific community has always questioned and debated new scientific concepts that someone has published. When Robert Hooke criticized some of his ideas put forth in *Opticks*, Newton became enraged and retreated into isolation for some time. In other incidences like this, he wrote nasty letters to colleagues who were his friends. Eventually, he suffered a nervous breakdown. In his later years, he published new editions of both *Principia* and *Opticks* in which he eliminated any references to the work of people he felt had wronged him. As president of the Royal Society, he used his position to attack his rivals. This irrational behavior caused him to be a solitary person with few good friends.

Isaac Newton in 1725 by Georges Vertue after John Vanderbank

Along with his scientific work, Newton was a dedicated public servant. He spent two terms as a Member of Parliament representing Cambridge. Later he moved to London to serve as Warden of the Royal Mint. In this position, he represented the monarch in collecting precious metals and printing money for England. In 1699, he was promoted to Master of the Royal Mint and served in this position until his death, 30 years later. As Master of the Mint, he supervised the recoining of English currency. During his time at the mint, Newton discovered and prosecuted many counterfeiters. He was elected president of the Royal Society in 1703 and was knighted by Queen Anne in 1705 for his service to the crown.

The third area he contributed to was theology. Newton's scientific investigations were a form of natural theology—that God is revealed in Creation. His theological work an extension of his scientific work. These studies reveal him as a devout man of faith, though he rejected some important doctrines of the faith. For example, he denied the Trinity and embraced Arianism, the belief that Jesus was a created being through whom all else was made—the most condemned heresy in the history of the church. He considered Christ "a divine mediator between God and man, who was subordinate to the Father who created him."[7] According to Newton, "The son in all things submits his will to the will of the father which could be unreasonable if he were equal to the father."[8] In other words, he denied the hypostatic union, the doctrine that Jesus was fully human and fully divine, two natures in one person.[9]

Newton wrote several books of biblical criticism, including *An Historical Account of Two Notable Corruptions of Scripture* (1754), in which he claims the early church did not believe in the Trinity, which was imposed on scripture later. He bases much of his argument on the infamous *Comma Johanneum*, a textual variant of 1 John 5:7 which inserts the phrase, "in heaven, the Father, the Word, and the Holy Spirit, and these three are one." Newton was right to reject the addition (which dates to the 10th c. and is clearly not original). However, the Trinitarian formula expressed by the *Comma* mirrors the theology of the early church as articulated in the Nicene Creed, which itself is a summary of what scripture teaches and is the standard of orthodoxy. Newton's ultimate motivation for rejecting the Trinity is his Arianism, not the *Comma*. He also wrote *Observations upon the Prophecies of Daniel, and the*

Apocalypse of St. John (1733) in which he dealt with predictions of the end of the world. Newton refused to publish either work during his lifetime, knowing some of the views would not only be controversial but cost him his positions as professor and Warden of the Mint (the penalty for denying the Trinity under the Blasphemy Act of 1697).

Along with denying the Trinity and the full divinity of Christ, Newton did not believe in the immortality of the soul or the existence of a personal devil. Although he was extremely serious about his religious beliefs, they were not always consistent with orthodoxy.

Newton carefully studied the Bible to try to find scientific information and hidden messages. The general belief at the time was that God periodically intervened in His creation to alter natural events. Newton strongly argued that God did not intervene in what was happening in the universe but rather had created everything using a rational design and universal principles that could be discovered by investigating the world.

Sir Isaac died in 1727 and had the honor of being the first scientist to be buried in Westminster Abbey. Although he was an imperfect man with many shortcomings, Sir Isaac Newton significantly changed how we think about the Creation in which we live. Most importantly he demonstrated how we can unify our understanding of areas of science such as the motion of our solar system and objects falling near the surface of the earth. This concept of unification was later used in other areas such as electricity and magnetism. We still use many of the theories and concepts he developed. In fact, everything we can observe of objects moving around us is still described using Newton's classical mechanics. Only the motion of small (atomic scale) objects or objects moving near the speed of light do not obey his predictions. He is remembered in history as one of the most influential people who ever lived, an outstanding scientist and a man of deep Christian faith—albeit unorthodox.

The tomb of Isaac Newton (By Javier Otero CC-BY-A3.0)

2.2 The Church in the Age of Enlightenment

Prior to the Enlightenment, belief in God or gods in some fashion was at the center of all cultures whether Christian, Jewish, Islamic, Hindu, Buddhist, or pagan. Christians took seriously the biblical refrain that "The fear of the LORD is the beginning of knowledge" (Proverbs 1:7 ESV).[10] For Christians, this referred to more than merely religious knowledge; it referred to *all* knowledge. John Calvin likened scripture to eyeglasses, an interpretive lens that enables us to clearly see God, and through him, the world and ourselves.[11] From Constantine through the Middle Ages, the church became increasingly entwined with civil institutions and rulers. By the beginning of the Renaissance in the 14th century, the Roman Catholic church had wielded tremendous political power for several centuries. The Christian worldview was inescapable in the West and knowledge was based primarily on authority.

When Constantinople was conquered by Islam in 1453, intellectuals fleeing the East brought a second infusion of classical Greek thought to Italy. This supplemented the rediscovery of the Greeks during the Reconquest of Spain 400 years earlier as well as Greek culture brought home by Crusaders. Greek influence flowered into a renewed interest in the arts, launching the Renaissance—a rebirth of human creativity and curiosity that lifted aesthetic achievement to unprecedented heights of excellence. For example, the new realism in how the human form was rendered stood in stark contrast to the more flat, stylized depictions in the art of the Middle Ages. It also represented a more profound philosophical shift from a God-centered worldview to one focused on human achievement and potential.

In the 1970s, theologian Francis Schaeffer used Michelangelo's *David* (1504) to demonstrate the new paradigm of thought in the Renaissance.[12] Although the assumption is that Michelangelo's statue is of the biblical King David preparing to fight Goliath, the only real connection to scripture is the name *David*. The biblical David was no older than fifteen when he fought Goliath. Michelangelo's *David* is clearly a fully grown young adult. David would never have fought in the nude, which was shameful in ancient Hebrew culture.[13] Most telling is Michelangelo's *David* is not circumcised and is therefore not part of the covenant God made with his people. He's not Jewish. The biblical David is a weak vessel through whom God's strength and covenant faithfulness to fight for Israel is revealed. Michelangelo's *David* is self-sufficient and proud; he can vanquish Goliath without God's help. Thus, Michelangelo used a biblical symbol to convey a humanist philosophy through the arts. The church's comprehensive hold on culture had cracked. Although religion remained important, a disenchantment with the supernatural and miracles began to change what it could explain.

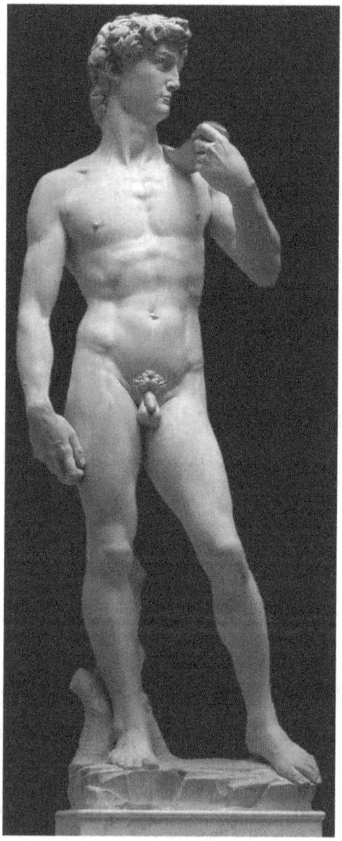

David by Michelangelo (GNU-FDL)

Little more than a decade after *David* emerged from marble, the crack was manifest in Germany with the Protestant Reformation, a concurrent cultural shift. After Martin Luther nailed his Ninety-five Theses to the church door in Wittenberg in 1517, they went viral thanks to the newly invented printing press. What Luther intended as church reform led to his excommunication, launching a new Christian tradition untethered from the centralized authority of the Pope. One of Luther's dangerous ideas (as Rome saw it) was to translate the Bible into the language of the people. No longer would Christians depend on the authority of the church to learn what it said. They could read it themselves. Luther did not deny the importance of tradition or authority, but he held to the principle of *Sola Scriptura*—the idea that all doctrine should be grounded in God's revelation. He agreed with Rome that this meant risking conflicting, idiosyncratic, and even heretical interpretations, but he broke with Rome in believing the benefit of having access to God's revealed word outweighed the risks.

Martin Luther in 1528 by Lucas Cranach

The fears came true, and the Protestant movement quickly splintered into a wide variety of denominations like light passing through Newton's prism. Inevitably, the convictions creating each new branch came into conflict with other branches, igniting the religious wars throughout Europe in the 16th and 17th centuries. Interestingly, the Protestant movement and its subdivisions brought with it an unintended consequence: the need for religious tolerance.

INDEX LIBRORVM
PROHIBITORVM,
CVM REGVLIS CONFECTIS
per Patres a Tridentina Synodo delectos,
auctoritate Sanctiss. D.N. Pij IIII,
Pont. Max. comprobatus.

VENETIIS, M. D. LXIIII.

The title page of the *Index Librorum Prohibitorum* from 1564

In response to the Reformation, the Catholic Church convened the Council of Trent (1545–1563). Among its many cannons was the creation of the *Index Librorum Prohibitorum*—the *Index of Prohibited Books.* Although several lists of banned books existed before Trent (including some made by Protestants), this list became the standard overseen by the Pope and used as a tool by the Inquisition.[14] At the time of Newton's death, the list included Protestant theologians (such as Luther, Calvin, and Melanchthon), philosophers (such as Hobbes, Descartes, Bacon, and Spinoza), and scientists (such as Galileo and Boyle). Although the *Index* was an attempt by the church to retain what control it could over the changing society, the very existence of the list was a measure of the increased emphasis on human reason and free-thinking (although the Reformers saw human reason unaided by scripture as a fatal move).

By the time Newton was born in 1642, the shift away from God at the center of life to humankind at the center was maturing into the Enlightenment. Although the freedom to express his original thinking was a product of his time, intellectual liberty did not extend to his most radical thoughts on religion. The Toleration Act of 1688 allowed nonconformist (non-Anglican) worship as long as it was within the bounds of Protestant orthodoxy. Heresy was explicitly prohibited by the statutes of the University of Cambridge even before the Blasphemy Act of 1697. As a result, Newton kept many of his convictions private and often wrote in code, only to be discovered and deciphered long after his death. His heretical views would have cost him his career if known, but they would possibly have cost him his life a hundred years earlier (as what happened to anti-Trinitarian Michael Servetus in 1553 at Geneva).

Newton's discoveries were nurtured by British Empiricism, the philosophical grounds for modern science. Early in the 17th century, Francis Bacon proposed a scientific methodology based on induction where experiments are designed and can be reproduced which results in theories that can be further confirmed as laws. The laws can then be the foundations on which new experiments can be built, resulting in a chain of progress toward more comprehensive knowledge of the world. However, this meant that scientific knowledge and religious knowledge are two different kinds of knowledge that don't interact.

John Locke in 1697 by Godfrey-Kneller

Philosopher John Locke (1632–1704), a contemporary of Newton's, proposed a unified method of knowing in *An Essay Concerning Human Understanding*, published in 1690. Locke believed people are born with minds like blank slates—they have no concepts *prior* to experience, only the ability to reason *from* experience. Experience generates simple ideas which are connected together by reason to create complex ideas. According to Locke, "Knowledge is the perception of the agreement or disagreement of two ideas."[15] With sense experience as the grounds of knowledge, both science and natural theology could be conducted with the same degree of certainty. This theory of knowledge didn't segregate religious and scientific knowledge. However, to make the theory work, God was no longer the foundation of knowledge, but the object of it using human experience as the foundation. It was a theory that could be embraced by Christians, skeptics, and the newly minted Deists with their mechanistic view of nature.

Thus, British Empiricism came into its own as Newton was putting the philosophy into practice. In fact, Locke and Newton corresponded with each other, exchanging ideas.

Even though Galileo and Newton both experienced a conflict between science and religion, it took different shapes. Neither scientist saw their discoveries as undermining their faith. Ironically, it was primarily the church that was threatened by the new knowledge. In the Renaissance, this was because the church had allied itself too closely with Aristotle and Ptolemy's cosmology. In the Enlightenment, the dramatic progress of science created an overly optimistic view of what science could explain, leading some to conceive of the universe as a machine that rendered God unnecessary. And the religious tolerance of the time, which allowed free-thinking and encouraged curiosity in science and philosophy, brought with it heterodoxy and heresy.

All the ingredients necessary for Newton's genius to take flight were perfectly mixed in the early Enlightenment. For those disturbed by his heretical views, the question is if that kind of revolutionary, individualistic thinker could be restrained in one field of study and still be groundbreaking in another. Or should we take the good with the bad and recognize that the kind of out-of-the-box mindset that brings such profound advances in science also casts a shadow by innovating in theology—a field where innovation is often tantamount to heresy? That he was deeply devoted to understanding the Bible and a committed Christian is beyond question. But despite how much he wrote on theology, he was first and foremost a scientist who discovered truths about the world we all benefit from. And that project revealed the creator God of the Bible he worshipped, even if he misunderstood him.

Sir Isaac Newton by H. Meyer

References and Notes

1. Newton, I., *Principia* (Amherst: Prometheus, 1995), 440.
2. Kant, I., *What is Enlightenment?*, 1784, included in *Foundations of the Metaphysics of Morals, and What is Enlightenment?* (New York: Liberal Arts Press, 1959), 85.
3. Janiak, A., Newton, I.. Correspondence with Richard Bentley [1692–3]. In: Janiak, A., ed. *Isaac Newton: Philosophical Writings.* Cambridge Texts in the History of Philosophy. (Cambridge: Cambridge

University Press, 2004), 94–105. DOI: https://doi.org/10.1017/CBO9780511809293.008.

4. Letter from Isaac Newton to Richard Bentley, 17 January 1692/3. https://www.newtonproject.ox.ac.uk/view/texts/diplomatic/THEM00255 (accessed 12.03.24).
5. Brewster, D., *Memoirs of the Life, Writings, and Discoveries of Sir Isaac Newton*, Volume II. (Edinburgh: Edmonston and Doublas, 1855), 331.
6. Turnbull, H.W., The Correspondence of Isaac Newton *Volume I 1661–1675* (Cambridge: Cambridge University Press, 1959), 416.
7. Westfall, R.S., *Never at Rest* (Cambridge: Cambridge University Press, 1980), 315.
8. Ibid, 316. Westfall quotes Newton's *Yahuda.* MS 15.25.
9. The hypostatic union was defined in 451 at the Council of Chalcedon, the Fourth Ecumenical Council, and is an essential doctrine of orthodoxy.
10. Cf. Proverbs 9:10; Psalm 111:10.
11. Calvin, J., *Institutes of the Christian Religion*, I.vi.1.
12. Schaeffer, F.A., *How Then Shall We Live?* (Old Tappan: Revell, 1976), 1–72.
13. Genesis 9:22–23.
14. The Index was abolished in 1966. Prior to that, some books on the list were eventually deemed safe for the faith and removed.
15. Locke, J., *An Essay Concerning Human Understanding*, IV.1.2.

Chapter 3

Michael Faraday (1791–1867): The Father of Electrical Engineering

For every house is built by someone, but the builder of all things is God.

Hebrews 3:4

"The book of nature which we have to read is written by the finger of God."

Michael Faraday

The Life and Faith of Michael Faraday

Imagine someone with musical talent who not only taught themselves how to play all the instruments in an orchestra but invented new ones. And at the same time, they composed symphonies of profound beauty using innovative structures and harmonic combinations that defied and expanded the current state of music theory, forever altering how music

Seven Saints of Science: An Introduction to the History of Conflict between Science and Religion
Richard C. Powell and Doug Powell

ISBN 978-981-5352-24-5 (Hardcover), 978-1-042-00201-6 (eBook)
www.jennystanford.com

DOI: 10.1201/9781042002016-3

Michael Faraday (© Doug Powell)

was made. And imagine they did it all without learning how to read music. Swap *music* for *science* and you have Michael Faraday.

Physicist Ernest Rutherford stated, "When we consider the magnitude and extent of his discoveries and their influence on the progress of science and of industry, there is no honour too great to pay to the memory of Faraday, one of the greatest scientific discoverers of all time."[2]

Faraday's scientific genius is why he is memorialized near Newton in Westminster Abbey's Scientists' Corner. Like

Newton, Faraday thought of himself as a natural philosopher who anchored his scientific inquiry in his Christian faith. Faraday believed, "The book of nature, which we have to read, is written by the finger of God."[3]

Faraday was born in Surrey (now part of London) to a poor family who belonged to a Christian sect called Sandemanians. The church had formed in Scotland in the 1730s based on the conviction that the kingdom of God Jesus taught so much about was a spiritual kingdom, not an earthly one. Therefore, they believed all national churches, such as the Church of Scotland, were unbiblical institutions. The sect did not object to Church of Scotland's Reformed theology, just the Presbyterian form of government and its recognition as the national church. In Scotland, the followers were called Glasites, after their founder. In England and other parts of the world where it spread, they were called Sandemanians, after the son-in-law of the founder who carried on the teachings.

The Sandemanians believed in a Congregational form of church government. Their most controversial tenet was not requiring pastors to have formal training. Taking the apostle Paul's charges to Timothy and Titus as normative (1 Timothy 3:1–13; Titus 1:5–9), they ordained godly men from within the congregations to be pastors, elders, and bishops. Other idiosyncratic practices included foot washing and refusal of worship with members of other denominations. They rejected the goal of accumulating wealth and believed private property could be claimed by the church to help the poor or the church itself. Admirably, they emphasized godly living over merely professing faith, which was exemplified throughout Faraday's life.

Faraday worked during the first half of Great Britain's Victorian Era, the period when it became the most powerful empire in the world. Queen Victoria's reign (1837–1901) was marked by unprecedented prosperity, an expanding economy, rich culture, and significant innovation. Much of this progress was driven by the Industrial Revolution that took place during this time. The invention of the steam engine and the development of machine tools stimulated the change from producing things by hand to manufacturing by machines. Factories no longer had to be located by running waterways to provide their power. New ways of manufacturing textiles and making iron were developed. This era of advances and new possibilities that began around the time of Michael Faraday's birth was the perfect environment for his creative brilliance to shine.

Queen Victoria in 1854

Faraday had little formal education and mostly taught himself after learning to read and write in Sunday school. As a young boy, he was apprenticed to a book binder for seven years, which gave him the opportunity to read many books, where he developed a fascination with chemistry, electricity, and other fields of science. In particular, Isaac Watts' *The Improvement of the Mind* had such a profound impact on his thought process that he thought no one should be without it.[4]

Sir Humphry Davy

At the end of his apprenticeship, a friend gave him tickets to attend lectures at the Royal Society, a national organization for advocating science. After taking detailed notes on the lectures of chemist Sir Humphry Davy, he sent them to Davy. Davy was so impressed that he supported Faraday's appointment as his Chemical Assistant at the Royal Institution (which focused on research) and began his scientific work in chemistry. He traveled around Europe with Davy visiting scientific colleagues, which helped him learn about many areas of science. This piqued his interest in doing his own research, and the Royal Institution gave him a laboratory, launching his research career.

Michael Faraday in the 1850s

Around this time, Faraday devoted his life to Christ and joined the Sandemanian church. He met his wife there, and he served as both deacon and elder in the congregation. They had no children.

Faraday made some outstanding contributions to chemistry, including the discovery of benzene and several other compounds. The field of metallurgy is based on his work on steel alloys. He experimented with liquifying gases, and in the process, invented rubber balloons. Faraday developed a new type of high-quality glass that was later used in telescopes. In addition, he wrote a manual of practical chemistry that was popular at the time.

Faraday was the first person to relate electricity to chemical bonding, combining his interests in electricity and chemistry. In addition, he studied electrolysis, the chemical changes that occur in a substance when electricity is passed through it. Faraday's laws of electrolysis are still used today and his work ushered in the field of nanoscience.

But his greatest contributions were in the field of electromagnetism. He studied the magnetism surrounding a wire carrying an electric current. This led him to develop the concept of the electromagnetic field, which changed how scientists thought about electricity and magnetism. Faraday's Law of Electromagnetic Induction shows that an electric current is induced in a conductive coil when it is in a varying magnetic field.

Interestingly, because of his poor education, Faraday could not use mathematics to describe this effect. Years later, James Clerk Maxwell described this effect mathematically, and it is still recognized as one of the basic laws of physics. Faraday's concepts of electromagnetic induction using rotating devices form the basis of electric motors and generators which, in time, facilitated the usefulness of electricity in technology. This drove the Industrial Revolution into the Age of Electricity, and because of this Faraday has been called the father of electrical engineering. He also demonstrated that electrical charges are only on the outside of a conductor so he could use a "Faraday Cage" to insulate an object inside from an electromagnetic field on its outside.

In his studies of magnetism, Faraday showed that some materials placed in a strong magnetic field produced a weak magnetic field in the opposite direction. These are called *diamagnetic* materials. He also showed that the polarization direction of a beam of light is rotated when it is passed through a magnetic field. This is known as the Faraday Effect and is used today in optical devices with laser sources. Because of Faraday's importance to the field of electromagnetics, the unit of capacitance is called the farad.

In time, Sir Humphry Davy came to resent the growth of Faraday's stature as a scientist. This caused them to have a major falling out, with Davy accusing Faraday of plagiarizing his work. The accusation was never substantiated and eventually Davy passed away, leaving Faraday to pursue his experiments in his lab at the Royal Institute.

In addition to his important research contributions, Faraday also provided intellectual consulting to various government and industrial groups. For example, he led an investigation of the problem with explosions in coal mines which resulted in enhanced safety for the miners. He improved England's maritime activities by using the high-quality glass he developed to make better lighthouses. He also helped determine ways to protect the bottom of ships from corrosion. However, when he was asked to help the government develop chemical weapons for the Crimean War, he refused due to his religious convictions.

Faraday believed in being a good steward of the environment. As part of his public service, he wrote about air pollution and water pollution in the Thames River. He was passionate about the education of young people and testified about problems with the British educational system before the Public School Commission. At the Royal Institution, he started the public "Christmas Lectures" to explain science to young people and the general public. These very popular lectures are still going on today. In addition, Faraday acted as an advisor for the Great Exhibition of 1851 in London and was a consultant to the National Gallery on cleaning and protecting its art.

Michael Faraday delivering a Christmas lecture

Faraday felt it was unbiblical to accept worldly rewards and pursue wealth. In a 1859 letter, he wrote, "I have always loved science more than money; and because my occupation is almost entirely personal, I cannot afford to get rich."[5] He was offered many honors that he declined due to his religious beliefs, such as a knighthood offered by Queen Victoria, as well as the presidency of the Royal Society. He did accept an honorary Doctor degree from the University of Oxford and was the first Fullerian Professor of Chemistry at the Royal Institution. He was also an elected foreign member of national scientific societies throughout the world. For his lifetime of service as a government consultant, he was awarded a house in Hampton Court, now known as the Faraday house.

Even as a child growing up, he built gadgets to show how something worked. His major contributions as a scientist were experimental demonstrations he developed in his laboratory. Because of his meager education, he lacked the mathematical expertise to express his results as equations. He left this to others, such as Maxwell, whose famous Maxwell's Equations of electromagnetic theory describe Faraday's experiments on electromagnetic induction.

Faraday worked long, stressful hours and was exposed to many different types of chemicals, such as mercury, in his research. This took a toll on his health, and he began to experience memory loss, vertigo, and bouts of depression. In 1839, he suffered a nervous breakdown but eventually returned to his investigations into electromagnetism at the Royal Institution. He retired to live in Hampton Court and died there in 1867. He rejected the honor of being buried in Westminster Abbey and was interred in Highgate Cemetery with a plaque placed in his honor at the Abbey.

Faraday wrote, "The natural works of God can never by any possibility come into contradiction with the higher things that belong to our future existence, ...still I do not think it at all necessary to tie the study of the natural sciences and religion together."[6] For him, the two issues were distinct things, though his work was grounded in his faith. He believed the natural laws were part of God's creation, and the secondary means through which he sustained it. Faraday's research was aimed at understanding these laws, and he was successful in doing this for electromagnetic fields. He said, "God has been pleased to work in his material creation by laws and the Creator governs his material works by definite laws resulting from the forces impressed on matter. The beauty of electricity is that it is under law. The laws of nature, as we understand them, are the foundations of our knowledge of natural things."[7]

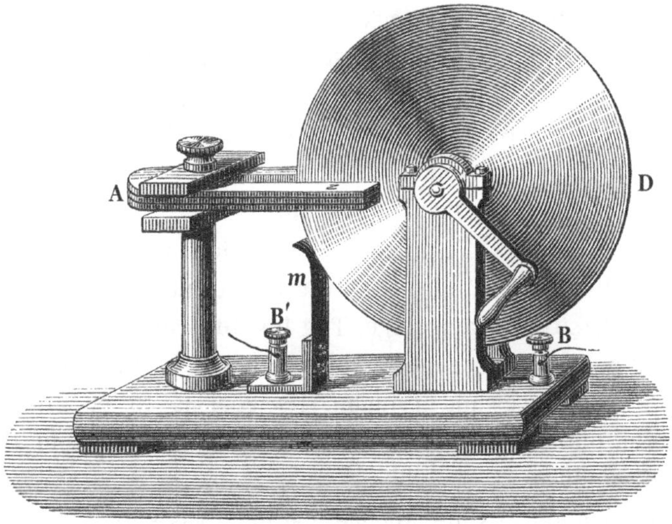

Faraday's disk generator

Two of Faraday's personal Bibles are currently in the archives of the University of Cambridge. The underlined passages and notes in the margins are the result of his daily study of scripture.[8] He used biblical quotes in his lectures such as Romans 1:20: "For since the creation of the world his invisible attributes—his eternal power and divine nature—

have been clearly seen, because they are understood through what has been made."

The most notated book in the Old Testament of Faraday's Bible is the book of Job, which emphasizes the finitude and fallibility of man and had a strong effect on Faraday. Admitting his own fallibility, he would publish the results of his scientific experiments but not argue in their favor to defend them. He felt it was up to other scientists to check their validity. This lack of self-promotion was quite unusual in the scientific community of his day.

As an experimental physicist and inventor, Michael Faraday was one of the most influential scientists of the Industrial Revolution, an amazing achievement given his humble beginnings and lack of early childhood education. In Victorian society, it was almost impossible for a poor person without formal education to rise to any degree of prominence, let alone in a field such as science. Faraday's success was even more remarkable considering his religious convictions which insisted on living a humble life.

In 2006, the Templeton Foundation provided funding to establish The Faraday Institute for Science and Religion at the University of Cambridge. The Institute supports academic research to advance the understanding of the interaction between science and religion. Its goal is to enhance the general public's appreciation for the compatibility of these two disciplines.

Throughout his life, Michael Faraday was known as a kind, humble person unconcerned with worldly honors, focused on pursuing his scientific experiments. He is a great example of success achieved by someone starting life in an under-privileged environment. His devotion to the Christian faith and his pursuit of scientific knowledge were both of great significance to him, and he saw no conflict between them. His faith shaped his character so that he comported himself with integrity and humility in both his life and his work.

Michael Faraday by J. Cook

Michael Faraday had the perfect intellect and skill to succeed during the Industrial Revolution. He devised and constructed insightful experiments demonstrating the scientific principles he studied, overcoming his lack of formal education. Although mathematics is the language of physics, his humble character shaped by his religious faith enabled him to let others describe his results mathematically. Although he didn't shy away from quoting scripture in his lectures, he did not use science as a proof of God's existence (even though he believed it was). Instead, Faraday primarily expressed his faith publicly by example—through his life and through serving as an officer in his church.

The Church in the Industrial Revolution

It is difficult to understate the profound impact the scientific discoveries of the Enlightenment had on the period following it. The development of the scientific method and the ever-increasing understanding of the laws of nature were more than academic achievements. They became tools used to harness nature in a way that ubiquitously improved the quality of life in every area of society. The application of science to create technology that increased productivity while easing burdens began a new age called the Industrial Revolution (1750–1850).

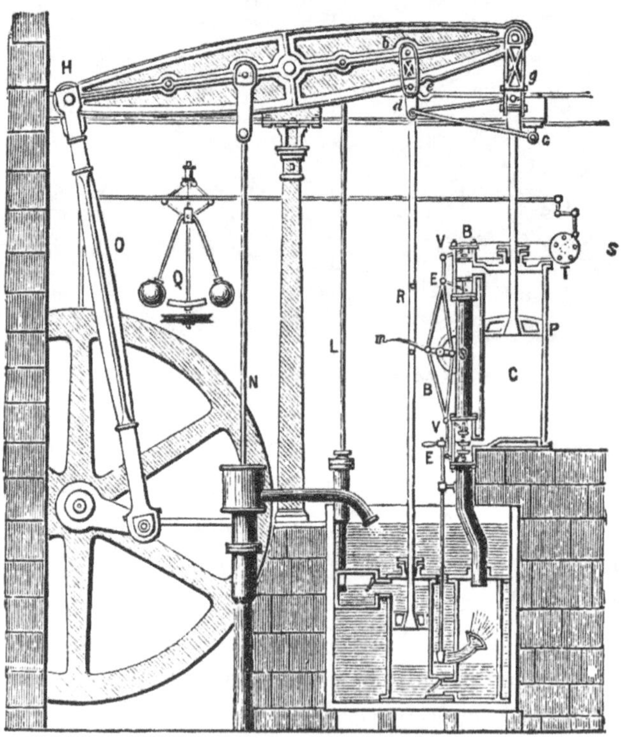

A diagram from 1780 of James Watt's steam engine

James Watt's improvements to the steam engine in the 1760s was a primary catalyst for change. It created a source of power far greater and more reliable than animal or human effort, and it did not need to be next to a water source to drive a wheel. Factories were built, centralizing and efficiently concentrating the work. New cities grew around factories, and cities such as London became hubs of factories. Factory work was higher paying and less physically demanding than working in fields, and the agrarian society became urbanized as people moved to the cities.

The development of cheap steel expanded the possibilities what could be made, and machine-made tools became more specialized for the tasks. New machines for weaving outpaced the production of thread, inspiring new spinning machines that could keep up with demand, transforming the textile industry. New chemical dyes changed the color palette for both high and low fashion. Further improvements to the steam engine made it small enough to mount on wheels and large enough to power ships, radically transforming transportation both on land and water, bringing more of the world within reach. Imported material increased what could be made, and exported goods found new markets.

Even darkness and time were impacted by new inventions. The manufacture of coal into coke created a flammable gas that was collected for use in lamps, pushing back the night through streetlamps. The lamps were also used in factories so that they could stay productive after the sun went down, creating work shifts regardless of the time of day.

The advances were a fulcrum in history, not only for Great Britain where it began, but for the entire world. Prior to the Industrial Revolution, the daily life and standard of living of the average person were almost indistinguishable from the lives of their parents, grandparents, great-grandparents, and ancestors. After the Industrial Revolution, progress was the norm. Each generation was expected to have advantages over the previous generation, an attitude baked into the modern mindset. History changed from a kind of moving walkway to an escalator.

The Industrial Revolution created unprecedented wealth and opportunity while raising the baseline of comfort across society. But the solutions manufactured by factories also created new problems. The conditions inside the factories too often made victims of the workers. Long hours, low wages, and unsafe environments were the cost of business, and the new machines operated without a conscience.

It was left to the church to call for reform and advocate for the poor working class. Charles Dickens' *A Christmas Carol*—published in 1843, toward the end of the Industrial Revolution—put the concern of the church in the mouth of Marley's Ghost, proclaiming, "Mankind was its business." And action wasn't restricted to the national churches, the Church of England and the Church of Scotland. It included the nonconformist churches that had proliferated under religious tolerance.

In addition to Methodists and Puritans, a number of denominations concerned with social reform descended from the Swiss Reformed movement of Anabaptists. Anabaptists held to separation of church and state, believed faith was more than an ascent to doctrines but must be lived out, and were strongly committed to mutual aid. Within the Anabaptists, a desire to return the church to its earliest form developed, the time of the Apostles. The primitive church movement believed the purest form of Christianity was before creeds and councils interpreted the teaching of the Apostles. The first sect in Scotland to split from the Church of Scotland because of a commitment to the primitive church were the Glasites or Sandemanians, the church to which Faraday belonged.

A Glasite meeting house in Perth, Scotland (by Kim Traynor CC-BY-SA 3.0)

The impetus for social engagement was, in part, due to a new approach to faith inspired by Romanticism. For some, the Enlightenment was overly focused on intellectual pursuits and ignored emotions and imagination—essential aspects of humanity. Romanticism was a reaction against the Enlightenment and the advent of the Industrial Revolution, meant to recover a place for feelings and intuition. It was primarily expressed in the arts through voices such as William Blake, Lord Byron, William Wordsworth, and Samuel Taylor Coleridge. Romantics were not anti-intellectual yet elevated the heart above the mind.

Romanticism found a place in theology as a reaction to a similar complaint. Too often, Christianity was treated as a set of propositions that were assented to and accompanied by baptism. Although neither of these were bad things, they did not result in a faith that impacted culture. In the eyes of many

Christians, they were necessary for faith, but not sufficient. What it lacked was action, a personal transformation that acted on those beliefs.

That there was truth in the critique could not be denied. The question was how to remedy the problem. John Wesley (1703–1791), a priest in the Church of England, developed an approach for cultivating a deeply personal and experiential faith after seeing the piety of a Moravian sect in London. His teachings inspired the Methodist movement, which eventually became its own denomination.

German theologian Friedrich Schleiermacher (1768–1834) went further, over-correcting against what he saw as cold, wooden faith. Schleiermacher wrote that rather than a set of doctrines, "The essence of religion consists in the feeling of an absolute dependence."[9] In making "God-consciousness" the grounds of faith, he traded God as an objective truth for the experience of God as a subjective feeling. The exchange eroded the ability to make claims of objective truth. On the other hand, emphasizing emotions also helped spur Christians into action to impact culture, giving them experiences where they lived out their faith through service.

An evangelical movement also developed within the Church of England that believed the church should be the leader in social reform. A small but highly influential group formed around William Wilberforce (1759–1833), a Minister of Parliament. In addition to successfully leading the decades-long charge against slavery, he and those in his group founded a number of aid societies that (among other things) cared for the poor, improved the conditions of prisons, advocated for the rights of workers, and condemned the abuse of animals.

The religious tolerance stemming from the Enlightenment resulted in several strains of Christianity that addressed the ills of society from both within and outside the national church. Many aid societies were not outreaches of an established church, but rather Christian organizations. Missionary societies took advantage of the advances in transportation and spread Christianity to parts of the world that had remained untouched by it. Other societies provided Bible translations for them, encouraging literacy.

William Wilberforce

The social engagement of Christians providentially made the faith relevant just when Christianity as a whole faced one of the most severe challenges in its history. In the shift from a God-centered world to a human-centered world, the Bible itself became an object to be studied and tested. German scholars in particular began to take the view that the Bible was a human book, not a divine one. They believed they found errors, contradictions, superstitions, myths, and textual corruption that undermined its authority. For Christians in the tradition of Schleiermacher, this was not a huge problem since feeling dependence on God was the essential aspect of faith. For traditional believers, however, the new claims required new responses. For example, biblical archaeology provided concrete evidence that corroborated some of the history contained in scripture. And the development of textual criticism developed a method for recovering the original text of scripture with a high degree of certainty based on the discovery and collation of every known manuscript of scripture prior to the invention of the printing press.

A flax-mill in Leeds in 1843

The advances of the Industrial Revolution generated an attitude of unrestrained optimism in the ability of science to solve problems that bettered society. Things that were unthinkable in one generation were common in the next. The place for miracles and the supernatural seemed to be diminished further with each new achievement. The Bible as the foundation of all knowledge seemingly became less and less tenable. Yet, none of the technological advances brought an alternate ground of morality or could account for the origins of existence. No invention amplified the conscience or provided purpose. And no machine could manufacture love.

Although the epicenter of the Industrial Revolution was England, important advances were made in other parts of Europe as well. In France, the research of Andre-Marie Ampère led to additional understanding in electromagnetism. His invention of the solenoid provided a way of creating a magnetic field using a wire coiled around a magnet. His understanding of the interaction between two wires carrying current came to be called Ampère's Law, and the unit of an electric current is named Ampère in his honor. Ampère was a devoted Roman Catholic who—like the protestant Faraday—embraced the science of the Enlightenment and saw a logical place for it within religious faith while rejecting its increasingly human-centered philosophy. He was also a mentor and friend of the Blessed Frédéric Ozanam, who lived with the Ampère family. Ozanam was the main founder of the Society of Saint Vincent de Paul Catholic charitable organization.

Another important non-British contributor to science during the Industrial Revolution was Italian astronomer Brother Angelo Secchi, S.J., a Jesuite priest. When the Italian revolution caused an upheaval in Italy, Secchi moved to England and then to the United States, where he studied at Georgetown University. While there, he met Mathew Fountain Murray, the first director of the U.S. Naval Observatory, who stimulated his interest in astronomy. He returned to Italy as director of the Observatory at the Pontifical Gregorian University. When the Kingdom of Italy took control of the observatory, Secchi was allowed to remain director despite refusing to sign a pledge of allegiance to the king. His excellent reputation as a scientist protected his position, and his unshakable faith and commitment to the church gave him a conviction earthly powers should not shake. Secci's innovations include being the first person to use a prism at the end of his telescope to study the spectrum of light coming from different stars. The data he obtained can be used to tell what gases are present in the star and its temperature. He studied and classified over 4,000 stars and is generally thought of as the father of astronomical spectroscopy.

Michael Faraday c.1860 by Millikan Gale

Michael Faraday was not only an invaluable contributor to the Industrial Revolution. He is an excellent example of someone who viewed his work in the context of faith. Where some saw a divide between science and religion, he planted a foot in each and found unity. Faraday understood that as great as human ability is, it is not autonomous. There is a holy ghost in the machine.

References and Notes

1. Faraday, M., Lectures on education, delivered at the Royal Institution of Great Britain (1854).
2. Rao, C.N.R., *Understanding Chemistry* (Cambridge: Cambridge Universtiy Press, 2000), 281.
3. Faraday, M., *Experimental Researches in Chemistry and Physics* (London: Richard Taylor and William Francis, 1859), 471
4. Correspondence of Faraday to Benjamin Abbot, quoted in Jones, Bence, *The Life and Letters of Michael Faraday*, vol. 1 (London: Longmans, Green, and Co, 1870), 210.
5. Correspondence of Farady to William Smith, Jan. 3, 1859, quoted in Jones, Bence, *The Life and Letters of Michael Faraday*, vol. 1 (London: Longmans, Green, and Co, 1870), 418.
6. Ibid, 1844 letter, 191–2.
7. Cantor, G., *Michael Faraday: Sandemanian and Scientist* (New York: St. Martins, 1991), 202.
8. Pratt, H.T., *Bulletin for the History of Chemistry* 11, 40, 1991.
9. Schleiermacher, F., *The Christian Faith*, 5.4–5.

Chapter 4

Lord Rayleigh (1842–1919): Country House Science

Great are the works of the L*ORD, studied by all who delight in them.*

Psalm 111:2 (ESV)

"I have never thought the materialist view possible, and I look to a power beyond what we see, and to a life in which we may at least hope to take part."[1]

Lord Rayleigh

4.1 Lord Rayleigh's Science and Faith

When a scientist makes an important discovery, it is often named for them. John Strutt has over *fifty* things in science named for him. Why haven't you heard of him? Because he was part of the British nobility and used the name of Lord Rayleigh after his Barony. The large variety of contributions

Seven Saints of Science: An Introduction to the History of Conflict between Science and Religion
Richard C. Powell and Doug Powell

ISBN 978-981-5352-24-5 (Hardcover), 978-1-042-00201-6 (eBook)
www.jennystanford.com

DOI: 10.1201/9781042002016-4

named after him include Rayleigh scattering, Rayleigh resolution criterion, Rayleigh waves, the Rayleigh number, Rayleigh flow, Rayleigh–Taylor instability, and the Rayleigh-Jeans law. All of these were named after John William Strutt, the leading scientist of the Victorian Era.

Lord Rayleigh (© Doug Powell)

Queen Victoria had the second-longest reign of any monarch in the history of England, ascending to the throne in 1837 and ruling until her death in 1901. Although this was a time of great prosperity in England, not everyone benefitted from it. English society was sharply divided into commoners and nobility. The nobility class was made up of titled peers and landed gentry. Titled peers were based on the system of an earlier era when the king assigned people to be his representatives in different regions of the realm. A ruling monarch can make new nobility by knighting someone as a reward for special service to the country. Some positions of nobility were passed on through inheritance while others were lifetime appointments. In the Victorian Era, the nobility was the privileged class of people with many benefits including opportunities for education and pursuing fields such as science and medicine. The commoners were much less fortunate. Much of what we know about life in the Victorian Era comes from the books of Charles Dickens (1812–1870), who wrote during that time. Today the major benefit for a member of the nobility in England is the ability to stand for election to the House of Lords.

John William Strutt was born in Essex, England, in 1842, five years into the Victorian Era. His father was the second Baron Rayleigh of Terling Place, a title in the British peerage established in 1821. Being from nobility, he had the opportunity to go to a top-ranked secondary school. He attended Harrow and Eton but in both cases had to withdraw because of poor health. Instead, he was educated at home with a tutor. He obtained his college education at Trinity College, Cambridge and after graduation was elected to a fellowship there.

In 1871, Strutt married Evelyn Balfour, whose brother became Prime Minister. They had three sons, and one of his sons (the fourth Baron Rayleigh) became an excellent physicist in his own right. In 1873, his father died, and he inherited the Barony of Rayleigh and was known as Lord Rayleigh for the rest of his scientific career. The title enabled him to participate in the House of Lords, but he spoke only when the political topic involved science.

Terling Place (by Chronicle – Alamy Stock Photo)

Lord Rayleigh was the Cavendish Professor of Physics at the University of Cambridge from 1879 to 1884 and was the university's chancellor from 1908 until his death in 1919. He was elected fellow of the Royal Society in 1873 and served as president from 1905 to 1908. Increasing the public's awareness of science was important to Rayleigh, and he acted as an advisor to the government in several positions. He chaired the Explosive Committee for the War Office and was president of the government's Advisory Committee on Aeronautics during World War I. He was the science advisor for Trinity House, which had the responsibility for coastal installations such as lighthouses and buoys. In addition, he advised the government in the areas of education, aviation, and the treasury.

Lord Rayleigh came from a family of wealthy landowners who had no background at all in science. As a member of the peerage, he was expected to take on social obligations which would leave no time for scientific research. However, he fell in love with experimental physics and had his younger brother take over responsibility for running their family agricultural business. He used some of his own resources to convert a stable block into his personal laboratory at his estate in Terling. At that time, there were very few laboratories in universities so it was not unusual for men interested in science to establish private laboratories if they could afford them. However, it was unusual for a peer of the realm to be active in any type of work and Lord Rayleigh was a good example of a hard-working peer in country house science. Many of his initial scientific investigations were done in this home laboratory with his wife as his chief assistant.[2,3] Lady Rayleigh had grown up in a household that valued science, and she proved to be extremely helpful. His sister-in-law and other family members also assisted at times.

Because both Lord and Lady Rayleigh were raised in evangelical families, a strong work ethic and the tradition of a wife supporting her husband were part of their belief. They also enjoyed travel, tennis, photography, and music.

In 1872, Lord Rayleigh came down with rheumatic fever. To recover, he and his wife took a long trip sailing down the Nile River in Egypt. However, this did not stop his scientific work. He used the time to write his classic work on acoustics, *The Theory of Sound,* which was published in 1877 and is still in use today.

Lord Rayleigh's time in his Terling laboratory between 1884 and 1908 was one of the most productive times for his research career because he was not encumbered by university administrative duties. Unfortunately, in 1908 his agriculture business began to have financial problems. He was mainly a dairy farmer and "Lord Rayleigh Dairies" in London provided outlets for his dairy products. Because of this economic downturn, he accepted the position of Chancellor of the University of Cambridge and served in this position until his death.

Cavendish Laboratory (by Spudgun67 CC-BY-SA 4.0)

During his time as chancellor, universities were establishing laboratories, institutionalizing science. The Cavendish Laboratories (comprising the physics department) were expanded after Lord Rayleigh generously donated his Nobel Prize money to the university in 1904. In addition to his administrative duties, he was able to continue his research. Lord Rayleigh was one of the most prolific scientists in history, publishing over 446 scientific papers. These were clearly written and showed that he had a deeper understanding of scientific concepts than most of his colleagues. In addition, he was a popular teacher and made sure that all his classes were open to women, which was unusual at that time.

Lord Rayleigh at work in his laboratory

Lord Rayleigh stands out both because of the number of important scientific contributions he made and because of the many different areas of physics to which he contributed. During his research on the density of gases—among the first inquiries in that area—he discovered a new gas he called *argon*. As a result, he was awarded the 1904 Nobel Prize in physics. It was the fourth year the prize was given to "those who have conferred the greatest benefit to mankind." Argon has found numerous uses but is especially useful as a laser.

One of his major areas of research was on the properties of waves, both optical and acoustical. An important result from his study of light scattering was explaining how light interacts with particles that are smaller than the light's wavelength. This "Rayleigh scattering" of sunlight explained that blue light was scattered more than red light, which is why the sky is blue. His work on waves also explained the "Whispering Gallery Modes" of sound waves in St. Paul's Cathedral. If you stand in a spot under one side of the dome, you can hear someone whispering a long distance away on the other side of the dome. In 1992, this concept of resonance was used to design a new type of microchip laser.[4] His work on the propagation of light waves on surfaces is now being used to describe solitons in optical fibers. And he established the criterion for the resolving power of optical instruments such as microscopes and telescopes.

Rayleigh also developed the theory describing the characteristics of radiation emitted by an object because of its temperature, called the Rayleigh-Jeans theory of blackbody radiation. The result turned out to be valid for long wavelengths but not short-wavelength radiation. This so-called "ultraviolet catastrophe" was one of the important problems explained by the early development of quantum mechanics.

Lord Rayleigh made equally important contributions to the physics of acoustics, fluid dynamics, and electromagnetism. His studies of fluid dynamics resulted in the Rayleigh number for convection, the theory of aerodynamic lift, and an understanding of friction and lubrication. In addition, he developed the experimental standards for the measurements of electrical units for current, resistance, and voltage.

Lord Rayleigh (by Elliott and Fry)

In 1902, he was honored as one of the first recipients (and first physicist) awarded the Order of Merit. King Edward VII established the honor to recognize people who made distinguished contributions to society. On receiving the honor, Rayleigh commented, "Any results that may be due to my researches were owing to the fact that it has been a pleasure for me to become a physicist."[5]

Although he did not mix science and religion and did not write about his religious views, Lord Rayleigh framed his scientific work with his Christian faith. He prefaced the third volume of his *Scientific Papers* (his collected works) with Psalm 111:2: "The works of the LORD are great, sought out of all them that have pleasure therein" (KJV). And he introduced the fifth volume with a verse from the Wisdom of Solomon: "Thou hast ordered all things in measure and number and weight" (KJV).[6] Rayleigh answered the question of God's existence, saying: "Suppose I discover a new mathematical theorem. Has it never been apprehended by any mind before? To me that is inconceivable."[7] This is essentially a restatement of Isaiah 55:9, which reads, "My ways higher than your ways, and my thoughts than your thoughts."

He also criticized atheism for failing to have the explanatory power to account for the world and our experience. In personal correspondence, he wrote, "I have never thought the materialist view possible, and I look to a power beyond what we see, and to a life in which we may at least hope to take part. What is more, I think that Christ and indeed other spiritually gifted men see further and truer than I do, and I wish to follow them as far as I can."[8] Rayleigh's humble and generous character is the fruit of his faith. In the same letter, he advised, "But the great thing is to pray, even if it be in a vague and inarticulate fashion. Surely we can ask a blessing on those we love and...that the Holy Spirit may in all things direct and rule our hearts."[9]

Lord Rayleigh was a committed Christian who had a major impact on science and saw no conflict between science and religion. He died in 1919 at his home in Essex and is buried in the graveyard at All Saints Church in Terling in Essex. Although he is buried in Terling, the wall of St. Andrew's Chapel in Westminster Abbey contains a memorial to "John William Strutt, 3rd Baron of Rayleigh...An unerring leader in the advancement of natural knowledge."

4.2 The Church in the Victorian Era

> *"It was the best of times, it was the worst of times, it was the age of wisdom, it was the age of foolishness, it was the epoch of belief, it was the epoch of incredulity, it was the season of Light, it was the season of Darkness..."*

Although the opening line of *A Tale of Two Cities* was how Charles Dickens described London and Paris in 1775—the overlap of the Enlightenment and Industrial Revolution—the dichotomies had only amplified by the time he wrote it in 1859, the midst of the Victorian Era. The Industrial Revolution brought not only increased productivity and new levels of wealth—the best of times, but a decrease in the conditions of the workers who tended it—the worst of times. The contrast spawned a distinctive Victorian frame of mind.

Jeremy Bentham (by Henry William Pickersgill)

Results-based scientific progress during the Industrial Revolution had an impact on philosophy. Jeremy Bentham (1748–1832) and John Stuart Mill (1806–1873) articulated a results-based view of morality that had no need for God. Utilitarianism equated human happiness with moral good, essentially sanctifying self-interest. In short, it argued the right thing to do was whatever did the most amount of good for the greatest number of people.

John Stuart Mill c1870

By the start of the Victorian Era, less than half of people regularly attended church in Great Britain.[10] And the first half of the 19th century saw Parliament relax and abolish statutory attendance laws. Yet despite the decline in numbers, Christianity remained a powerful voice in society, particularly through Evangelicalism. The stress on personal faith by Evangelicals carried with it a commitment to family, self-improvement, a strong work ethic, sympathy for the poor, and social reform.

Interestingly, Utilitarianism and Evangelicalism found common cause on several points. Both were grounded in a morality that relied on individual liberty (a product of tolerance) for actions that resulted in the betterment of society as a whole. The intersection of sacred and secular was a moralism that characterized the era. It was the conscience that found the sympathy to address the kinds of abuses that created unsustainable conditions such as documented by Dickens in *Hard Times* (1854):

> *It was a town of red brick, or of brick that would have been red if the smoke and ashes had allowed it; but as matters stood, it was a town of unnatural red and black like the painted face of a savage. It was a town of machinery and tall chimneys, out of which interminable serpents of smoke trailed themselves for ever and ever, and never got uncoiled. It had a black canal in it, and a river that ran purple with ill-smelling dye, and vast piles of building full of windows where there was a rattling and a trembling all day long, and where the piston of the steam-engine worked monotonously up and down, like the head of an elephant in a state of melancholy madness.*[11]

The strict class system of that time left the working class—the largest class—with little voice to bring attention to the abusive conditions they suffered. The upper class—the smallest class—was comprised of titled peers. They were the principal landowners and had a political voice. The middle class was made up of successful businessmen such as factory owners, and professionals such as doctors and lawyers. The Industrial Revolution had produced a greatly expanded middle class by the Victorian Era. Their success provided them with funds

A factory town in Wales c.1830

to buy land, which gave the men the right to vote. Dickens himself was counted among them. His authentic voice came from his family's personal experience with debtor's prison and workhouses before his uncommon rise out of the working class.

Parliament had enacted reforms through the 1833 Factory Act and 1834 Poor Act, but more needed to be done. Although three-quarters of Christians were still at least nominally Anglican, the voluntary nature of nonconformist denominations was well-suited to activism and helped lead to the creation of numerous benevolent aid societies independent of the church. These societies offered solutions to improve the lot of the poor working class through philanthropy, virtues shared by Christians and Utilitarians.

The freedom necessary for religious tolerance eventually redounded to the Church of England through the Oxford Movement. The Oxford Movement believed the church risked being seen as a function of the government and wanted to re-establish a clearer distinction between church and state.

As the British Empire expanded across the globe, missionaries (mostly nonconformist) accompanied it and went beyond its boundaries. The Victorian ethos could be found in India, Africa, and throughout the territories of the Commonwealth of Nations. At home, nonconformist churches proliferated and many Anglican churches were restored. In addition to worship, the church buildings were used to serve the community such as by hosting labor union meetings or serving food to the poor. Religion's moral compass remained steady, keeping it relevant despite a decline in Christianity and the new challenges to it.

F.C. Baur, a theologian at the University of Tübingen in Germany, eroded trust in the New Testament by applying the Hegelian philosophy of thesis/antithesis to the conflict between followers of Peter and Paul. He proposed a synthesis that united the early church in the second century. This community then produced the books of the New Testament, dating them long after the age of the Apostles, making them historically unreliable. Baur's influence led to a critical view of the text of the Bible. New critical methods were developed that claimed to show both the Old and New Testaments were products of editors who created and compiled the biblical books long after their traditional authors had died. At a time when anti-supernatural sentiment was on the rise, the new Higher Criticism found a ready audience.

F.C. Baur c.1855–1860 (by F. Brandseph)

In response, Christian scholars such as Constantine Tischendorf began a quest for the oldest Greek New Testament manuscripts. The discipline of textual criticism was developed to recover its original text. The discovery of codices Sinaiticus,

Vaticanus, Alexandrinus, Bezae, and Ephraemi in particular led to an increasing ability to recover the original text and produce more reliable translations of scripture.

Title page of Ussher's Annals from 1658

The turn of the 19th century saw the advent of geology, which became another point of conflict with the church. Throughout church history, theologians with a high view of scripture understood the six days of creation in Genesis in different ways. Even among the early church fathers, some interpreted them literally,[12] and others figuratively.[13] And among the ones who held a figurative view of the days, some also held Creation was 6,000 years old.[14] There was no monolithic view in the church on the days of Creation other than God as the ultimate cause and acting agent. In 1648, Bishop James Ussher of Ireland published his influential *The Annals of the World,* where he used the genealogies in the Old Testament to calculate the date of creation as October 22, 4004 BC.[15] But in the late 1700s, after observing the oldest rocks in the Earth's crust were sedimentary, James Hutton explained the Earth's surface was gradually developing at an extremely slow rate. He also worked under the theory of Uniformitarianism, the idea that the natural forces we see in the world have been the same throughout history. Sedimentary rock was produced using erosion under pressure and heat, a process that required a dramatically longer time than a literal reading of the Bible allowed. In 1833, Charles Lyell published the first book on geology, introducing the new field of science to a wide audience while further developing Hutton's ideas. Additional discoveries and investigation into the Earth's strata confirmed their theory, indicating the Earth was millions of years old.

Logically, either the interpretation of the scientific data was wrong, or the interpretation of the Bible as portraying the Earth as thousands of years old was wrong, or both interpretations were wrong. Princeton theologian B.B. Warfield—a strong defender of the inspiration of the Bible—accepted the science and criticized the method of using pre-Abrahamic genealogies to construct a chronology, calling it "entirely factitious," with "no solid basis."[16] But despite the increasing weight of evidence supporting the ancient age of the Earth, many Christians saw geological science as conflicting with the Bible because of Ussher's influence and a literalistic interpretation of Genesis 1. Young Earth Creationists continue this debate to this day, basing chronology on genealogies in scripture while reinterpreting the scientific data through a theory of catastrophism, where a sudden catastrophe (such as in a literal Creation day or Noah's Flood in particular) can account for the features of the Earth.[17]

Charles Darwin in 1854

And even stronger challenge came in 1859 with the publication of Darwin's *On the Origin of Species*. Darwin showed the traits of animals and plants evolved over time through a process of natural selection that make them better suited for their environment. His work theorized that all different species could have evolved from one prototype of life. The new discoveries in geology provided a far greater age for the

Earth that accommodated Evolution. Darwin believed his theory of Evolution was compatible with belief in God, although he himself explicitly rejected Christianity after formulating his theory and considered himself agnostic. He didn't publicly share his views on religion, restricting his comments to personal correspondence,[18] but more vocal skeptics (such as Thomas Huxley) used the theory of Evolution to argue that God was not needed to explain creation. This argument between creationists and evolutionists continues today. Some within the church (such as the BioLogos Foundation) accept Evolution as one method God uses as the Creator of everything, some believe the theory is incompatible with scripture, and some believe the theory lacks evidence to explain origins.

This era was the first time that the general public took interest in the arguments between the church and science rather than letting the two institutions play out the conflict largely outside the public sphere. This made the age of the Earth and the Theory of Evolution more than factual disputes. Two pillars of authority in society—science and religion—were in apparent conflict. Although some had room to accept the theories into their views of scripture, many could not reconcile them with their interpretations. As a result, a culture war was born which continues into contemporary society.

Another interesting development toward the end of the Victorian Age was the growth in atheism and bold philosophical expressions of it. In 1843, philosopher Karl Marx wrote, "Man makes religion, religion does not make man.... This state and this society produce religion, which is an inverted consciousness of the world, because they are an inverted world."[19] In 1882, philosopher Friedreich Nietzsche (1844–1900) went further, declaring, "God is dead! God remains dead! And we have killed him!"[20] Both Marx and Nietzsche resonated with an atheistic view of Evolution and left a profound political and social impact on the 20th century.

The Great Exhibition of 1851 (by H.F. Talbot)

The Great Exhibition of 1851—the first world's fair—was held in London, a sign of the new importance of science in society and the awareness of it by the public. As the century progressed and more advances were made, different people reacted in different ways. For some, the new developments showed that human beings were progressing and that the problems of the future could be solved with appealing to religion. For others—such as Lord Rayliegh—science was a worldly supplement to eternal truths revealed by God. B.B. Warfield described the end of the Victorian Era, writing in 1889, "Science has not broken with religion. But a large number of the scientific thinkers of our generation have."[21]

References and Notes

1. Strutt, R.J., Baron Rayleigh, *John William Strutt: Third Baron Rayleigh, O.M., F.R.S., Sometime President of the Royal Society and Chancellor of the University of Cambridge* (London: E. Arnold & Company, 1924), 361.

2. Opitz, D.L., Country-House Science in Late Victorian Britain, Ph.D. Thesis, University of Minnesota, 2004.
3. Gavin, W., *Ninety Years of Family Farming: The Story of Lord Rayleigh's and Strutt & Parker Farms* (London: Hutchinson, 1967), 37.
4. McCall, S.L., Levi, A.F.J., Slusher, R.E., Pearton, S.J., Logan, R.A., *Applied Physics Letters*, 60, 289 (1992).
5. Court Circular, published in *The Times of London*, August 9, 1902, p. 6; *London Gazette*, September 2, 1902, 5679.
6. A book in the Apocrypha, which is included in the Bible used in the Church of England.
7. Strutt, R.J., Baron Rayleigh, *John William Strutt: Third Baron Rayleigh, O.M., F.R.S., Sometime President of the Royal Society and Chancellor of the University of Cambridge* (London: E. Arnold & Company, 1924), 361.
8. Ibid.
9. Ibid.
10. Field, C.D., Counting Religion in England and Wales: The Long Eighteenth Century, c. 1680–c. 1840. *The Journal of Ecclesiastical History*. 2012;63(4):693–720. doi:10.1017/S0022046911002533
11. Dickens, C., *Hard Times*, chapter 5. Parts of this passage were used to illustrate the same point in Shelley, Bruce, *Church History in Plain Language* (Nashville: Thomas Nelson, 1982, 1995), 405.
12. Cf. Basil the Great, *Hexaemeron* II.viii; Lactantius, *The Divine Institutes* VII.xiv; Theophilus of Antioch, *Autolycus* III.xxvi.
13. Cf. Justin Martyr, *Dialogue with Trypho* LXXXI; Iranaeus, *Against Heresies* V.xxiii.2; Augustin, *The City of God* XI.vi.
14. Cf. Augustin, *The City of God* XII.10.
15. Ussher uses the Julian calendar date. The Gregorian calendar date is September 21.
16. Warfield, B.B., *The Princeton Theological Review*, IX, 1911, 2.
17. Cf. Answers in Genesis is a well-known advocate of catastrophism. https://answersingenesis.org/geology/.
18. Cf. Darwin, C., *Life and Letters of Charles Darwin, The*, Darwin, Francis, ed. (London: D. Appleton and Company, 1898). A chapter VIII is dedicated to his thoughts on religion.

19. Marx, K., A Criticism of the Hegelian Philosophy of Right, in *The Collected Works of Karl Marx* (Hastings: Delphi Classics, 2016), 14.

20. Nietzsche, F., *The Complete Works of Friedrich Nietzsche,* vol X, Levy, O., ed. (Edinburgh & London: T.N. Foulis, 1910), 168.

21. Warfield, B.B., Homiletic Review, Jan 1889, pp. 9–16. Included in *Warfield, B.B., Selected Shorter Writings of Benjamin B. Warfield* vol. II (Nutley, NJ: Presbyterian and Reformed, 1973), 132.

Chapter 5

Albert Einstein (1879–1955): The Genius and "Cosmic Religious Feeling"

The dice are thrown into the lap, but their every decision is from the LORD.

Proverbs 16:33

"God does not play dice."

Albert Einstein[1]

Albert Einstein (© Doug Powell)

Seven Saints of Science: An Introduction to the History of Conflict between Science and Religion
Richard C. Powell and Doug Powell

ISBN 978-981-5352-24-5 (Hardcover), 978-1-042-00201-6 (eBook)
www.jennystanford.com

DOI: 10.1201/9781042002016-5

Einstein's Science and Faith

Have you ever called someone really smart an "Einstein"? Who was this man whose name is used to mean *genius*? Albert Einstein was the greatest physicist who ever lived because he made so many unique contributions to science.

Albert Einstein in 1921 (by F. Schmutzer)

As the world changed from the 19th to the 20th century, physics changed from "classical physics" to "modern physics." Up through the 19th century, physicists tried to explain how everything worked in the everyday world we see around us. They thought they could explain everything with intuitive thinking. However, there were a few unsolved problems. One was the "ultraviolet catastrophe" where the classical

Rayleigh-Jeans theory of the frequency distribution of radiation emitted by a hot object predicted an infinite value at short wavelengths (an impossibility). Another was the photoelectric effect, where the characteristics of electrons knocked off the surface of metal by a beam of light could not be explained by the classical theory of the energy of a light wave. The solution to these problems required the development of new concepts in physics.

During the first quarter of the 20th century, physicists changed from intuitive thinking to creative thinking, a paradigm shift that established modern physics. The new approach changed the focus of research to explain the extremes of the universe: the properties of extremely small particles, how objects act when moving at extremely fast speeds, how extremely large objects interacted with each other, and what happens to matter at extremely low temperatures. Much of the physics studied in these different conditions involved the interaction of light with matter. In general, modern physics can be divided into two areas called "quantum mechanics" and "relativity." Together they revolutionized the field of physics.

This period of science was the perfect time for a free-thinking, creative genius like Albert Einstein to come onto the scene. Einstein played an important role in the development of both quantum mechanics and relativity because the concepts he developed were not small steps forward but dramatic advances into areas that had never before been considered. They were so innovative and complicated that the validity of some of his theories is still being tested today.

One of Einstein's major contributions was to reformulate the theory of electromagnetic radiation, the way we think about light. Instead of a wave, he pictured light as a beam of tiny particles called *photons*. His theory was useful in explaining the properties of electrons knocked off the surface of a metal by a beam of light. This "photoelectric effect" was an important problem that classical physics had not been able to explain. Now we think of light as having a dual nature, sometimes acting like a wave and sometimes acting like a beam of particles.

In Einstein's reformulated theory of electromagnetic radiation, he included a new process of an atom emitting light by "stimulated emission." This seemed to be a minor extension of the theory of light. However, over 40 years later physicists finally learned how to use this process to make a laser (Light Amplification by Stimulated Emission of Radiation). The laser is one of the most important inventions of the 20th century and is now used in many areas, including communication, medicine, industry, military, and entertainment. Every time you scan a bar code at the store, you are using a laser.

Einstein thought a lot about how the results of a scientific experiment might be different depending on where the scientist was when a measurement was made. By then, physicists had determined that no object could be accelerated to speeds greater than the speed of light, which they designated as *c*. However, no one had explained why this was true or how objects behave if they travel at a speed very close to *c*. Einstein's unique insight discovered the answers to the questions. He began by stating two things must be true: first, the speed of light is the same no matter where you are when you make a measurement; second, the laws of physics are the same no matter where you are when you make a measurement. Then he performed a thought experiment of two people making a measurement of the same event occurring in a spaceship moving by the Earth near the speed of light. One person is traveling on the spaceship where the event occurs, and the other is looking at the event through a telescope on the Earth. In his theory of special relativity, Einstein showed that the person on the Earth would measure a slower time and a shorter distance in the direction of motion than a person traveling on the spaceship would measure for the same event. These strange differences have actually been proven to be true. In addition, Einstein's theory predicts that an object's mass increases when it approaches the speed of light and would be infinitely large at *c*. According to classical physics, no force could be great enough to accelerate an object with an infinite mass and therefore its speed can never reach *c*.

Albert Einstein at his office at the University of Berlin in 1920

When the property of an atom—such as the absorption or emission of light—is measured experimentally, you don't get a single, specific answer but rather a range of answers. The reason for this is that atoms are always moving, and observers will measure shorter frequencies for the atoms moving toward them and longer frequencies for atoms moving away from them. Since the motion of the atom restricts the accuracy of the measurement (in this case frequency) it would be helpful to make the measurement with the atom at rest. The motion of an atom is related to its temperature. Although it is not possible to cool something to absolute zero temperature[2] with normal refrigeration techniques, it is possible to stop the motion of an atom using lasers. An "atom trap" can be formed by hitting an atom with laser beams from all sides to suspend it in space which makes it at absolute zero temperature.[2] According to a theory developed by Satyendra Nath Bose and Einstein, when a group of atoms that have the property of integer spin angular momentum are at very low temperatures they condense into a new state of matter with special quantum mechanical characteristics. This so-called "Bose–Einstein Condensation" was achieved experimentally using laser traps at both the University of Colorado, Boulder, and MIT in the 1990s, confirming the theory.

A horseshoe Einstein ring (by ESA-Hubble & NASA)

Einstein also worked on reformulating the concepts of gravity which is called "General Relativity." Classical physics used Newton's ideas of gravity as a force acting at a distance between two objects. Instead, Einstein pictured gravity as an indentation in four-dimensional space-time. In classical physics, an object had to have mass to be affected by gravity. However, in Einstein's formulation something like a beam of light with no mass would still be affected by gravity. Astronomers have now observed the bending of light emitted by distant stars as it goes around massive stars or galaxies to reach the Earth. Other predictions of General Relativity are still being tested. For example, Einstein's theory predicts the existence of gravitational waves, which were finally detected in 2015.

One of Einstein's most important contributions was showing the equivalence of mass and energy resulting in what has been called the greatest scientific equation ever written, $E = mc^2$. This equation is fundamental for understanding the nuclear processes of fission and fusion which are important for generating energy, nuclear medicine, and making weapons.

In 1921, he was awarded the Nobel Prize for using the quantum theory of light to explain how light knocked electrons out of metals (photoelectric effect), which could not be understood by classical physics. All of Einstein's contributions revealed a unique, original way of thinking. When asked how he did this, Einstein said, "I am enough of the artist to draw freely upon my imagination. Imagination is more important than knowledge. Knowledge is limited. Imagination encircles the world."[3]

Einstein applied the same original thinking that brought dramatic change to our scientific understanding to his life and religion. However, outside of science, he was in general a very normal human being. Albert Einstein was born in Ulm, Germany in 1879 and died of an abdominal aortic aneurysm in Princeton, New Jersey in 1955 at the age of 76. Albert and his younger sister were raised in a middle-class Jewish family. Although he was instructed in the Talmud as a child, the family did not practice the faith seriously. Their father moved to Milan, Italy, for work after the factory he ran failed, leaving Albert to finish school in Munich. However, Einstein was unhappy there and didn't want to get drafted into the German military, so he moved to Italy to be with his parents. As a child, he was fascinated with a compass, wondering how invisible forces could make a needle move, a topic he worked on for the rest of his life. He also became enamored with a geometry book, which spurred his interest in mathematics.

Outside of science, Einstein developed two lifelong interests: sailing on his sailboat and playing violin. He sailed all his life and found pleasure and relaxation in his boat after struggling for hours on complex physics problems. Because of his absentmindedness, he was not a great sailor, and many times had to be towed in after running his boat aground or even capsizing.[4] His other joy was playing the violin, which he learned from his mother as a young child. His favorite composers were Mozart and Bach. He once said, "If I were not a physicist, I would probably be a musician. I often think in music. I live my daydreams in music. I see my life in terms of music.... I cannot tell if I would have done any creative work of importance in music, but I do know that I get most joy in life out of my violin."[5]

Einstein finished his high school education in Switzerland and then earned a bachelor's degree at the Swiss Federal Institute of Technology. During graduate school at the University of Zürich he had many friends, including Milena Maric, a fellow physics student from Serbia. They married in 1903 and had a daughter and two sons. Unfortunately, like many dedicated professionals, Einstein was so focused on his work that he was not a great family man. In 1919, he and Milena divorced, and Einstein married Elsa Lowenthal, one of his cousins. Einstein had little to do with his children from his first marriage, but Elsa brought two daughters with her from a previous marriage, and he raised them as his stepdaughters.

In 1905, Einstein was awarded a Ph.D. from the University of Zürich. Ironically, his professors were reluctant to write strong recommendations which made it difficult for him to get an academic position. Instead, Einstein began his career working at the Swiss patent office. This was actually a great job for him since he could complete his required work quickly and spend the rest of his time thinking about the physics problems that interested him. 1905 was a special year for Einstein. That was the year he published four important papers that revolutionized physics. One explained the photoelectric effect, one explained Brownian motion, the third introduced the concepts of special relativity, and the fourth showed the equivalence of mass and energy. The profound impact of these papers allowed him to move on to an academic career doing research and teaching at several universities in Germany, Austria, and Switzerland before ending up at the Institute for Advanced Studies at Princeton in 1933.

His main motivation for leaving Europe for the United States was the rise of Hitler and Naziism in Germany. He was an ardent pacifist and could not abide the mistreatment of any fellow human beings. The Nazi regime took over control of the scientific community in Germany. They called Einstein's work "Jewish science," and like everyone with his ethnic background in Germany, he was threatened with bodily harm. Once he left Europe, Einstein never returned to Germany.

Einstein spent the rest of his life at Princeton trying to develop a unified theory of physics that included both gravity and electromagnetism. His focus on this work caused him to become somewhat isolated from the mainstream physics community. He passed away before he was able to finish his unified field theory.

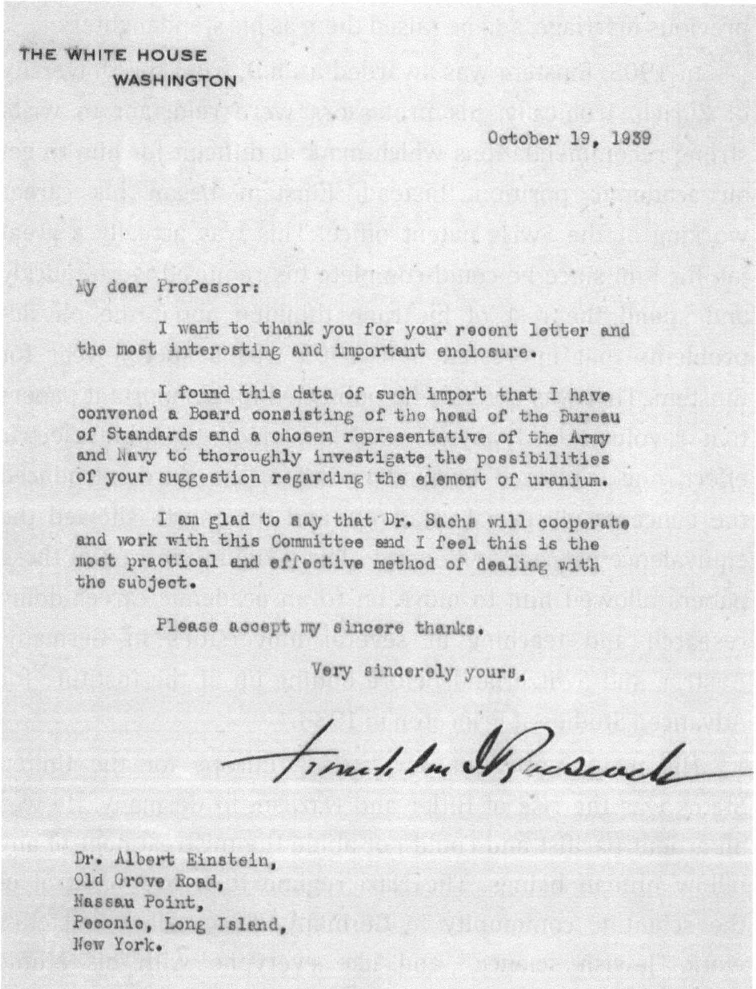

THE WHITE HOUSE
WASHINGTON

October 19, 1939

My dear Professor:

I want to thank you for your recent letter and the most interesting and important enclosure.

I found this data of such import that I have convened a Board consisting of the head of the Bureau of Standards and a chosen representative of the Army and Navy to thoroughly investigate the possibilities of your suggestion regarding the element of uranium.

I am glad to say that Dr. Sachs will cooperate and work with this Committee and I feel this is the most practical and effective method of dealing with the subject.

Please accept my sincere thanks.

Very sincerely yours,

Franklin D Roosevelt

Dr. Albert Einstein,
Old Grove Road,
Nassau Point,
Peconic, Long Island,
New York.

Roosevelt's 1939 letter to Einstein

Einstein was outspoken about the problems of prejudice in the United States but felt the situation was much better in this country than in Europe. In August 1939, he was asked to sign a letter to President Roosevelt describing the work the scientific community (including in Germany) was doing in the area of nuclear fission to obtain a chain reaction.[6] He described its potential for being the basis of a weapon of mass destruction. Roosevelt took this letter very seriously and initiated the Manhattan Project, leading the United States and its allies to build an atomic bomb before the axis powers could build one. Although the A-bomb was responsible for bringing about a quick end to World War II, it is ironic that such a strong pacifist as Einstein stimulated its development. Except for the initial letter, Einstein played no direct role in the Manhattan Project.

Einstein in 1947

Einstein's thoughts about religion were as unique and complex as his thoughts about science. He admired Jesus as a moral teacher, just as he admired the Buddha and other religious figures, saying, "No one can read the Gospels without feeling the actual presence of Jesus.... No man can deny the fact that Jesus existed, nor that his sayings are beautiful."[7] He also engaged religious views outside his tradition, such as his interaction with Indian philosopher Rabindranath Tagore. Although he was ethnically Jewish and supported establishing a Jewish state, he did not practice the faith, going so far as to declare, "I am not a Jew."[8] He also declined an offer to become the president of Israel when it was formed. He believed, "The Jewish religion like all others is an incarnation of the most childish superstitions."[9]

He rejected the biblical description of God as personal, calling it anthropomorphic. According to Einstein, "human fantasy created gods in man's own image,"[10] and "The word god is for me nothing more than the expression and product of human weaknesses."[11] Yet Einstein frequently spoke about God and used anthropomorphic language to do it. "We all dance to a mysterious tune, and the piper who plays this melody from an inscrutable distance—whatever name we give him—Creative Force, or God—escapes all book knowledge."[12] He spoke of God as a "giver," believing God made him a humanitarian,[13] and that, "The source of all scientific advancement is the God-given curiosity of the toiling experimenter and the constructive fantasy of the technical inventor."[14] And one his favorite aphorisms was, "God does not play dice with the world."[15]

At the same time, he defined God in impersonal terms, often using equivocal explanations that conflated God and the universe. "The whole of nature is life, and life, as I observe it, rejects a God resembling man. I like to experience the universe as one harmonious whole. Every cell has life."[16] He said, "I see a clock, but I cannot envision the clockmaker. The human mind is unable to conceive of the four dimensions, so how can it conceive of a God, before whom a thousand years and a thousand dimensions are as one?"[17] Yet after confessing the incomprehensibility of this impersonal God, he writes, "I meet (God) every day in the harmonious laws which govern the universe.... My religion is cosmic, and my God is too universal to concern himself with the intentions of every human being.... My God speaks to me through his laws."[18]

Einstein in 1946

Einstein described his "Cosmic Religion" as belonging to a higher order of thinking than most people could attain. He viewed religion as having three stages of development. The first stage is based on fear, focused on survival. The second stage is based on society and focused on morality, with rewards and punishments. The third stage can only be reached by "exceptionally gifted individuals or especially noble communities."[19] This is where, "the individual feels the vanity of human desires and aims, and the nobility and marvelous order which are revealed in nature and in the world of thought."[20] For Einstein, "the most important function of art and of science is to arouse and keep alive this feeling in those who are receptive."[21] In fact, "the cosmic religious experience is the strongest and the noblest driving force behind scientific research."[22]

Einstein's view of God as the universe—an impersonal force who enigmatically chooses, acts, and is rational—led to an ethic that is paradoxical at best. He believed, "Everything is determined, the beginning as well as the end, by forces over which we have no control."[23] But he was aware of the moral and ethical implications of his view. "I am a determinist. As such, I do not believe in free will.... Practically, I am, nevertheless, compelled to act as if freedom of the will existed. If I wish to live in a civilized community, I must act as if man is a responsible being."[24] Despite being one of the elite enlightened people who understand Cosmic Religion, he must regress to the second level of religion to ground his ethics. His determinism can't condemn the evils of Hitler's Germany, yet his grounding of morality in his second stage of religion justified his moving to the United States to escape it.

In the end, Einstein's esoteric Cosmic Religion is a scientific pantheism expressed in gnostic terms, where knowledge of the true religion is only accessible to an elite few. He wrote, "A contemporary has said, not unjustly, that in this materialistic age of ours the serious scientific workers are the only profoundly religious people."[26] Einstein's system inadvertently makes himself a kind of secular priest, leading some to describe him as an atheist (which he explicitly rejected) and others as a theist (which he heavily qualified). If Cosmic Religion has a creed, it is "Science without religion is lame, religion without science is blind."[26]

It is clear that the greatest physicist who ever lived was a profoundly spiritual man even though he rejected all traditional religions. His significant contributions to science were done within the framework of religion, albeit an idiosyncratic one. From a Christian standpoint, Einstein's extraordinary insight is an example of the doctrine of Common Grace, which allowed him to explore general revelation. For Einstein, "True religion has been ennobled and made more profound by scientific knowledge."[27] Or, as the Apostle Paul wrote in Romans 1:20, "For since the creation of the world his invisible attributes—his eternal power and divine nature—have been clearly seen, because they are understood through what has been made." Interestingly, in Romans 1:18 Paul also claims that people suppress the truth in unrighteousness. Einstein's idiosyncratic religion is a testament to both verses. Despite his genius and imagination, Einstein could not avoid appealing to God. His profound insights into the universe were ultimately insufficient to fully account for it.

5.1 The Church in the Age of New Physics

The unprecedented, exponential increase of scientific progress during the 19th century transformed the world, charting a course of limitless improvement. In 1800, crossing the Atlantic from England to New York took 5–8 weeks depending on winds for the sailing ships. In 1900, the same trip took five days on a steam-powered ocean liner.[28] In 1800, a package sent from New York to San Francisco took six months by wagon service. In 1900, transcontinental rail service delivered it in five days.[29] In 1800, a long-distance communication was by hand-carried letter. In 1900, the same message was telegraphed almost instantly—even across an ocean. In 1800, candles and oil lamps created dim oases of light. In 1900, electricity illuminated city streets and businesses, and began to light homes, overthrowing the tyranny of night.

Chicago's World's Columbian Exhibition of 1893 showcased the progress made since the Great Exhibition of 1851 in London. A moving walkway and the electric dishwasher gave new glimpses of a task-saving future. In particular, the electricity provided by Tesla's alternating current, manufactured by Westinghouse, proved the deciding factor in adopting its safer power over Edison's direct current method.

World's Columbian Exposition in Chicago of 1893

The scientific advances generated levels of optimism in human ability that had not been seen before. The contrast between the beginning and end of the century reinvented everyday life for the better almost beyond recognition. And in the face of the strongest challenges of science to the Bible (and religion in general), as well as critical historical attacks on scripture, a reappraisal of human beings occurred organically. Rather than seeing human beings through the doctrine of Original Sin—inheriting a fallen nature as descendants of Adam—people began to see themselves as fundamentally good. The proof was in the progress.

Progressivism became a philosophy to cultural leaders such as H.G. Wells. As a futurist, Wells envisioned what the world would be like at the end of the 20th century based on the progress of the 19th century. He saw a world without nations, war, disease, criminals, marriage, or religion. Technology would enable people to work for only a few years or even months in their lives, mainly to tend the devices that would take over the drudgery of work.[30] Although not all progressives had such a clearly defined view of what the future could be, his assessment of human ability was shared by many. In his *Outline of History*, he wrote, "We believe in the power of reason and in the increasing good-will in men."[31]

A French town destroyed in World War I

But like the Titanic—the ship that God couldn't sink—struck an iceberg in 1912, the naïve optimism of progressivism struck its own iceberg two years later with the outbreak of World War I. The Industrial Revolution created new weapons that made killing more efficient and brutal. Airplanes gave war new dimensions, and chemicals turned the air itself into a weapon. Human creativity and inventiveness turned on itself like a snake swallowing its own tail. Evil was undeniably real, and just as undeniably came from people. And the evil was on a grander scale than ever before. But Christians united in common cause in a just war had the vocabulary for the moral atrocities where the progressives did not.

The dissolution of the Ottoman Empire and the establishment of British and French mandates that followed allowed progress to spread more quickly into parts of the world that had remained relatively untouched by the technology and knowledge developed in the West. Although the European restructuring of the Ottoman Empire into the Middle East was flawed and sowed the seeds of future conflict, the advances imported from the West benefitted many people, giving them new opportunities and freedoms.

In the West, the vast casualties revived an interest in religion, especially Spiritualism. People craved knowledge of the afterlife as a source of comfort as they mourned the loss of fathers, sons, and brothers. Spirit mediums claimed seances were avenues to the other side where they could contact loved ones who had passed away. However, the movement was based on a pair of sisters in upstate New York who had played a practical joke on their parents by making knocks and clicking sounds without being detected. They claimed the sounds were made by departed spirits in response to questions asked of them. Within a few years, unscrupulous magicians had improved the techniques and exploited those who were grieving or spiritually hungry. The charade even grew to attract Sir Arthur Conan Doyle, creator of Sherlock Holmes and Spiritualism's most famous evangelist, who had lost a son to the war. Doyle believed the phenomena could be scientifically studied, a view that for a time was shared by Lord Rayleigh. By the mid-20s, seances had become so popular—and controversial—that Congress held hearings as it deliberated about whether to regulate them. Houdini himself crusaded against it, testifying before Congress and devoting part of every show of his last tour to exposing the unscrupulous methods of magicians who exploited the grieving by pretending to be mediums.

Although progressives were deeply shaken by the Great War, Wells and others retained enough faith in humanity to press even further to offer a solution: one world government. He wrote, "The world perishes unless sovereignty is merged and nationality subordinated. And for that the minds of men must first be prepared by experience and knowledge and thought.... In no other fashion is a secure world peace conceivable."[32] For progressives, human knowledge was salvation.

Thoughtful Christian theologians, such as Princeton Seminary's J. Gresham Machen, criticized progressivism, showing it did not belong to the future, but to the past. Machen wrote,

> *"Paganism is that view of life which finds the highest goal of human existence in the healthy and harmonious and joyous development of existing human faculties.... Paganism is optimistic with regard to unaided human nature, whereas Christianity is the religion of the broken heart.... (Paganism ignores) the disturbing fact of sin...Christianity means that sin is faced once for all, and then is cast, by the grace of God, forever into the depths of the sea."*[33]

Machen's insight was not unique. Many of Christianity's brightest minds collaborated on a series of essays beginning in 1910 aimed at answering challenges to the faith and defending the historic doctrines defined by the Nicene Creed. The inspiration of scripture, the virgin birth, miracles, the bodily Resurrection of Jesus, and his death as atonement for sin were among the doctrines that were affirmed. The collected essays were called *The Fundamentals.* Apologists, theologians, and pastors on both sides of the Atlantic contributed to the work, such as B.B. Warfield from Princeton, and James Orr from Trinity College in Glasgow. The result birthed a new movement called *Fundamentalism.* Unlike the anti-intellectual reputation the movement soon came to have, Fundamentalism began as a scholarly project that was highly informed, rigorously rational, and devoid of emotional appeals. Unfortunately, it hit its own iceberg.

Clarence Darrow (left) and William Jennings Bryan (right) during the Scopes Trial in 1925 (by Brown Brothers)

In 1925, a Dayton, Tennessee high school teacher named Scopes was prosecuted for teaching the theory of evolution, a violation of state law. The resulting trial attracted worldwide coverage. Although Scopes was convicted,[34] his lawyer, Clarence Darrow, won the court of public opinion. He successfully belittled the arguments of the well-intentioned but ill-prepared William Jennings Bryant. As a result, the trial accelerated a withdrawal from culture by Fundamentalists.

One aspect of the Scopes trial that is often overlooked is that the textbook Scopes used to teach the theory of evolution, *A Civic Biology*, used the theory to justify eugenics. The eugenics movement sought to improve the human race through selective breeding of "brilliant men and women" and sterilization or forced celibacy of the "low and degenerate races."[35] It was the basis of Hitler's attempt to develop an Aryan master race and had influential advocates in the United States such as Charles Lindbergh and Alexander Graham Bell.[36]

The conflict between science and religion that had grown more heated in the last quarter of the 19th century and first quarter of the 20th century became a culture war. Of the 182 colleges and universities in the United States in 1860, only seven were not founded as Christian institutions.[37] By the time of the Scopes trial in 1925, many of the schools had drifted far from their faith-based roots. A strong faction of the Fundamentalists responded by starting Bible colleges where the worldview often took precedence over the quality of education. In 1937, the secularization compromised even Princeton Seminary, leading to half its faculty—including Machen—quitting *en masse* to found Westminster Theological Seminary in Philadelphia in order to preserve the "Old Princeton" theology.

By the end of the 19th century, physics embraced a causal, deterministic philosophy. All changes had a specific cause and the future could be predicted with enough knowledge about the present. If you knew the current position and motion of a particle at any point in time, the laws of physics could accurately predict the position and motion at any future point in time. The new physics shattered this view. Quantum mechanics, relativity, and astrophysics produced radical new concepts that some philosophers attempted to apply to areas outside of physics, such as religion.

Just as Lyell and Hutton's geological data dramatically altered the understanding of the age of the Earth, Edwin Hubble's astronomical observations changed the understanding of the age of the universe. And the evidence carried startling implications with it. In the 1920s, Hubble studied light from

Edwin Hubble.

the most distant stars he could view through his telescope and discovered the wavelengths were elongated. He inferred the stars emitting the light were moving away from the Earth, and that the "Red Shift" was explained by the Doppler Effect. And since every star he observed showed the same Red Shift, it meant the universe was expanding, not in a steady state as previously thought. Logically, that indicated the universe began at a single point. A few years before Hubble, Georges H.J.E. Lemaitre, a Roman Catholic priest-astronomer, was the first person to recognize that the recession of galaxies was evidence of an expanding universe. Extrapolating this expansion back in time led him to propose a theory of the origin of the universe that later became known as the Big Bang Theory. Hubble and Lemaitre's observations had obvious theological implications. Some transcendent entity would have to exist to initiate the bang, bringing something from literally nothing. Yet, at the same time, some Christians didn't appreciate the observation's compatibility with Genesis 1:1, instead rejecting the evidence because the universe was far older than allowed in their interpretation of the days of Genesis.

Einstein and Robert Oppenheimer

The new physics created paradoxical and counterintuitive concepts such as duality, uncertainty, exclusion, discreteness, action at a distance, and the equivalence of the basic parts of creation. Duality says that at the atomic scale, everything exists as both a wave and a particle. The wave-like nature of a quantum system allows it to initiate action in a separate part of the system that has been moved some distance away. Particles in quantum systems can only exist in states having specific properties and not with any arbitrary set of properties.

All other particles in the system are excluded from having the same set of properties. The accuracy of measuring the exact value of one property is limited by how accurately the value of a related property is known (such as position and momentum or time and energy). Finally, the two fundamental properties of creation—mass and energy—were shown to be equivalent. The conversion between mass and energy was demonstrated by nuclear fusion and fission processes.

None of these strange properties of the new physics describing quantum systems on an atomic scale could be observed in normal everyday life. Some of these properties, such as action at a distance, were difficult even for some leading physicists like Einstein to believe. However, this has now been experimentally proven along with the other quantum properties.

The philosophy of the new physics moved away from determinism. It also gave an important role to the observer making measurements since this helped influence the outcome of the observation. The action of the observer determines the reality of nature at the time he is observing it.

In addition, the atomic scale of quantum mechanics exhibited an apparent chaos instead of an order that followed natural laws. Some philosophers tried to use this to argue that there was no divine creator, that the universe appeared randomly out of nothing. However, others argued that the quantum world could be observed only on the atomic scale with experimental conditions of extremely low temperatures and sophisticated equipment. The macroscopic world still obeyed the well-known laws of physics.

Buddhists, such as the Dalai Lama, easily accepted new quantum concepts into their religion, where the goal of enlightenment is to empty oneself of desire and want, which is the cause of suffering. He wrote, "If on the quantum level, matter is revealed to be less solid and definable than it appears, then it seems to me that science is coming closer to the Buddhist contemplative insights of emptiness and interdependence."[38] The concept of *emptiness* in Buddhism means that nothing has inherent existence. It exists only because you observe it. This is similar to the quantum idea that matter exists as a set of waves until it is observed—a way of thinking that was attractive to Einstein.

The new physics took science beyond strict empiricism to a place where observation did not mean seeing directly. Ironically, the language of indirect evidence providing data for what could not be seen was analogous to religious arguments—a parallel not lost on C.S. Lewis, a contemporary of the era of new physics. In *The Screwtape Letters* (1943), Lewis wrote from the perspective of an experienced demon to his young nephew, giving advice on how to best keep his assigned human from faith in God. Screwtape says, "Do not attempt to use science (I mean, the real sciences) as a defense against Christianity. They will positively encourage him to think about realities he can't touch and see."[39]

The early 20th century, when modern physics concepts of quantum mechanics and relativity were first developed, was the perfect time for an imaginative, free-thinker like Albert Einstein to work. His innovative theories were useful in explaining puzzling experimental observations by postulating certain basic principles, then reasoning how the physics would work. His rationalistic philosophy was instrumental in the development of his cosmic religious beliefs. He believed humankind comes close to God through understanding nature and thoughtful meditation—admittedly idiosyncratic religious beliefs throughout his life.

References and Notes

1. Hoffmann, B., *Albert Einstein* (London: Hart-Davis, MacGibbon, 1973), 193.

2. Absolute 0 is 0 on the kelvin scale, which is equivalent to -273.15° celsius and -459.57° farenheit.
3. Viereck, G.S., "What life means to Einstein," *Saturday Evening Post,* Oct. 26, 1929, 117.
4. Kilgannon, C., Recalling Albert Einstein, who was not much of a genius when it came to sailing, *New York Times*, July 22, 2007.
5. Viereck, G.S., "What life means to Einstein," *Saturday Evening Post,* Oct. 26, 1929, 113.
6. The letter is preserved at the Franklin D. Roosevelt Presidential Library and Museum, Hyde Park, NY.
7. Viereck, G.S., "What life means to Einstein," *Saturday Evening Post,* Oct. 26, 1929, 117.
8. Ibid.
9. Letter from Einstein to Eric Gutkind, Jan 3, 1954. Quoted in "Childish superstition: Einstein's letter makes view of religion relatively clear," *The Guardian,* May 13, 2008. https://www.theguardian.com/science/2008/may/12/peopleinscience.religion (accessed 12.24.24).
10. Einstein, A., "Science and Religion," paper presented at the Conference on Science, Philosophy and Religion, 1941.
11. Letter from Einstein to Eric Gutkind, Jan 3, 1954. Quoted in "Childish superstition: Einstein's letter makes view of religion relatively clear," *The Guardian,* May 13, 2008.
12. Hermanns, William, *Einstein and the Poet* (Boston: Branden Press, 1983), ebook location 323.
13. Ibid, ebook location 2302.
14. Einstein, A., *Einstein on Cosmic Religion and Other Opinions & Aphorisms* (Mineola: Dover, 1931, 2009), 93.
15. Hermanns, ebook location 1269.
16. Hermanns, ebook location 1393.
17. Einstein, A., *Einstein on Cosmic Religion and Other Opinions & Aphorisms* (Mineola: Dover, 1931, 2009), 102.
18. Hermanns, ebook location 1915.
19. Einstein, A., *Einstein on Cosmic Religion and Other Opinions & Aphorisms* (Mineola: Dover, 1931, 2009), 47.

20. Ibid, 48.

21. Ibid, 50.

22. Ibid, 52.

23. Viereck, G.S., "What life means to Einstein," *Saturday Evening Post,* Oct. 26, 1929, 117.

24. Ibid., 114.

25. Einstein's Three "Religious Geniuses,'" *New York Times Magazine,* November 9, 1930, pp.1–4. https://www.nytimes.com/1930/11/09/archives/einsteins-three-religious-geniuses-what-democritus-st-francis-and.html?searchResultPosition=1 (accessed 12.24.24).

26. Einstein, A., "Science and Religion," paper presented at the Conference on Science, Philosophy and Religion, 1941.

27. Ibid.

28. https://transportgeography.org/contents/chapter1/transportation-and-space/world-circumnavigation-days/; https://transportgeography.org/contents/chapter1/emergence-of-mechanized-transportation-systems/liner-transatlantic-crossing-time/; https://transportgeography.org/contents/chapter1/emergence-of-mechanized-transportation-systems/pre-industrial-travel-modes/ (accessed 12.26.24).

29. https://transportgeography.org/contents/chapter1/transportation-and-space/time-mail-delivery-new-york-san-francisco/; https://transportgeography.org/contents/chapter1/transportation-and-space/world-circumnavigation-days/; https://transportgeography.org/contents/chapter1/emergence-of-mechanized-transportation-systems/ (accessed 12.26.24).

30. Cf. Wells, H.G., *The Outline of History,* vol. II (London: George Newnes Limited, 1920), 757.

31. Ibid, 751.

32. Ibid, 752.

33. Machen, J. Gresham, *Christianity and Liberalism* (New York: MacMillan, 1923), 65–66.

34. The conviction was overturned on appeal.

35. Hunter, George William, *A Civic Biology* (New York: American Book Company, 1914), 263.

36. For an excellent in-depth look at scientific abuses, see *When Science Goes Wrong* by Br. Guy Consolmagno, SJ and Christopher M Garney (Paulist Press, New York, 1989).

37. Schmidt, Alvin J., *Under the Influence: How Christianity Transformed Civilization* (Grand Rapids: Zondervan, 2001), 190

38. Dalai Lama, *The Universe in a Single Atom* (New York: Harmony, 2005), Kindle iOS version, location 564.

39. Lewis, C.S., *The Screwtape Letters* (London: Centennial Press, 1942), 14.

Chapter 6

Werner Heisenberg (1901–1976): The Atomic Bomb that Never Was

"I made the earth; I created the people who live on it. It was me—my hands stretched out the sky. I give orders to all the heavenly lights."

Isaiah 45:12

"Physics is the reflection on the divine ideas of Creation, therefore physics is in divine service."

Werner Heisenberg[1]

Werner Heisenberg (© Doug Powell)

Seven Saints of Science: An Introduction to the History of Conflict between Science and Religion
Richard C. Powell and Doug Powell

ISBN 978-981-5352-24-5 (Hardcover), 978-1-042-00201-6 (eBook)
www.jennystanford.com

DOI: 10.1201/9781042002016-6

Heisenberg's Science and Faith

What would have happened if the Nazis had invented the atomic bomb before the United States did? The world would be a much different place than it is today. Given the expertise German scientists had in nuclear physics, it was not only a real possibility, but it was a goal they had and a race they lost. So, why did the Nazis fail?

Otto Hahn in 1913

At the start of World War II, Germany was at the forefront of scientific progress. It had some of the best scientists in the world, especially those knowledgeable about nuclear physics. Nuclear fission was discovered in 1938 in Germany by Otto Hahn. However, when Hitler rose to power in the

1930s, the Nazis vowed to replace "Jewish Science" with "Aryan Science."[2] Nineteen Nobel laureates and many other scientists fled Germany during this time, taking their expertise to England and the United States. Sir Ian Jacob, Winston Churchill's Military Assistant Secretary, said the allies won the war, "because our German scientists were better than their German scientists."[3]

The Germans had all the ingredients to create the first functioning atomic bomb: the knowledge, the resources, and the motive. This super-weapon based on nuclear fission would change not only the outcome of the war but the balance of power in the world for decades to come. To counter Germany's efforts, the allies started the Manhattan Project, which successfully built the first atomic bomb. Although it was not used until three months after Germany surrendered, Japan surrendered within days after seeing the massive, indiscriminate devastation at Hiroshima and Nagasaki. Although many German scientists left the country during the Nazi regime, Werner Heisenberg, one of the key scientists in the development of quantum physics, stayed and became the leader of their atomic bomb project. Ironically, his chief contribution was making sure the project failed.

Werner Heisenberg was born in Germany in 1901. His father was a university professor, and Werner attended excellent schools in Munich, eventually getting his Ph.D. in physics from the University of Munich. He worked as an assistant to Max Born in Göttingen, and then to Niels Bohr in Copenhagen. Heisenberg taught at a variety of different universities and eventually was appointed Professor of Physics at the University of Berlin and Director of the Kaiser Wilhelm Institute for Physics. After the war, he started what is now the Max Planck Institute for Physics in Göttingen, which he later moved to the University of Munich.

Heisenberg was one of the key physicists in developing quantum mechanics, including a matrix approach for mathematically expressing quantum mechanical systems and events. This was a very different way of viewing quantum mechanics compared to the wave theory that had been developed. He is best known for the Heisenberg Uncertainty Principle, which says the values of two related variables (such as position and momentum, or time and energy) cannot be known exactly at the same time. Although not important in everyday life, this concept is extremely important on the atomic scale. In 1932, Heisenberg was awarded the Nobel Prize for Physics for his work on quantum theory.

Himmler (left) and Hitler (right) inspecting Nazi troops

Heisenberg stayed in Germany throughout World War II to defend the honor of German science of Nazi domination. The Nazis wanted to purge German science of the influence of Jewish scientists, especially in theoretical physics. But Heisenberg refused to become a member of the Nazi party, opposed the appointment of Nazis to prominent scientific positions, and refused to sign a pledge of support for Hitler. In retaliation, Heinrich Himmler, Chief of German Police, appointed a committee of three Nazi physicists to investigate Heisenberg. The Nazis attacked him in the newspapers, calling him an "Aryan Jew." However, Heisenberg's mother was friends with Himmler's mother, and the two women convinced Himmler to stop persecuting Heisenberg because the Germans needed his physics expertise. Himmler relented and wrote a letter to Heisenberg cautioning him to keep his scientific work separate from his political beliefs.

Heisenberg in 1927

Despite his resistance to the Nazis, Heisenberg agreed to work on Germany's nuclear energy project. Under his leadership, the German scientists made progress on the development of nuclear reactors but never came close to making a bomb. After the war, Heisenberg revealed that he and other members of his small team steered the research in this direction on purpose, intentionally slow-walking the development of the bomb in part by overestimating in their reports how much it would cost and how long it would take. Heisenberg said the tactic, "enabled the physicists to influence further developments. If it were impossible to produce atomic bombs, this problem would not have arisen, but if they were easily produced, the physicists would have been unable to prevent their manufacture. This situation gave the physicists at that time decisive influence on further developments, since they could argue with the government that atomic bombs would probably not be available during the course of the war."[4]

Shortly after World War II ended, Yale physicist Henry Margenau hosted Heisenberg in his home. Margenau wrote, "Among most of my colleagues there existed the belief that he was a Nazi, who had worked for Hitler doing research on the atomic bomb which, if he had succeeded, might have destroyed our country and our lives. As a guest at my house he relieved himself by confiding in me and telling me exactly what the motives for his research had been. While he did work on nuclear physics and was aware of its implications he and his group worked on the theoretical issues and he avoided everything that might have enabled Hitler to use nuclear power as a weapon."[5] In other words, Heisenberg and his team reported enough progress to keep them in charge of the project, but without making the real progress necessary to achieve the goal. They appeared to make progress while actually restraining it.

In addition to scientific research, Heisenberg was an accomplished pianist. He also served as President of the Alexander von Humboldt Foundation, which sponsored students from throughout the world to work in Germany for a time. When he died of cancer in 1976, his colleagues walked from the Institute of Physics in Munich to his home and lit a candle at his door.

Werner Heisenberg in 1933

Heisenberg's Christian faith grounded his work and the moral convictions that guided it. He was a lifelong practicing Lutheran, but did not accept his faith merely because he was raised in the church. He cultivated his faith reading widely on religion, including Eastern religions, which contained teachings similar to concepts in quantum mechanics. Heisenberg was also interested in both modern and Greek philosophy and was particularly influenced by Plato.

Heisenberg believed his physics research was nothing less than understanding God's creation and was essentially a religious activity. He wrote, "Galileo argued that nature, God's second book, is written in mathematical letters...Kepler is even more explicit in his work on world harmony; he says: God created the world in accordance with his ideas of creation. These ideas are the pure archetypal forms which Plato termed *Ideas*, and they can be understood by man as mathematical constructs. They can be understood by Man, because Man was created in the spiritual image of God. Physics is reflection on the divine ideas of Creation, therefore physics is divine service."[6]

Science and religion were not only compatible for Heisenberg; they were complementary "facets of reality."[7] In discussing the beliefs about science and religion which he held in common with his colleague Max Planck, he explained, "Science deals with the objective, material world. It invites us to make accurate statements about objective reality and to grasp its interconnections. Religion, on the other hand, deals with the world of values. It considers what ought to be or what we ought to do, not what is. In science we are concerned to discover what is true or false; in religion with what is good or evil, noble or base. Science is the basis of technology, religion the basis of ethics. In short, the conflict between the two, which has been raging since the eighteenth century, seems founded on a misunderstanding, or, more precisely, on a confusion of the images and parables of religion with scientific statements."[8]

The complementary relationship between science and religion was a theme he often returned to. His thinking provides important context for his work during World War II. For example, he wrote, "[I]t has repeatedly been claimed that scientific truth cannot be reconciled with the religious interpretation of the world. Although I am now convinced that scientific truth is unassailable in its own field, I have never found it possible to dismiss the content of religious thinking as simply part of an outmoded phase in the consciousness of mankind, a part we shall have to give up from now on. Thus in the course of my life I have repeatedly been compelled to ponder on the relationship of these two regions of thought, for I have never been able to doubt the reality of that to which they point."[9]

Heisenberg in 1955 (by Vincenzo108 CC-BY-SA 4.0)

Without religion, Heisenberg understood there could be no coherent basis for morals and ethics in society. A materialist view of the world where science was the only means of knowledge could not provide a moral framework. "Where no guiding ideals are left to point the way, the scale of values disappears and with it the meaning of our deeds and sufferings, and at the end can lie only negation and despair. Religion is therefore the foundation of ethics, and ethics the presupposition of life."[10] This reflects his belief in the Lutheran doctrine of the three uses of the Law of God. The first is to restrain sin; the second to reveal sin and therefore our need of a savior; the third to guide our actions away from sin.[11]

Werner Heisenberg's Christian faith guided his thoughts and actions throughout his life, including how he viewed science. He believed modern physics should be a unifying influence on the world, decreasing political and cultural tensions by bettering the world through science. Conversely, he also understood that the same knowledge could be used for the creation of new weapons that threatened a unified culture. No other scientist with his convictions held a position within Nazi Germany who could surreptitiously prevent scientific advances from being used to create the first atomic bomb.

The Church during World War II

World War II was a time of great technological advances that supported the war effort, which, in turn, led to important civilian spin-offs. Two significant developments were radar (which led to the invention of television), and nuclear weapons (which stimulated the development of atomic energy and nuclear medicine). Other examples included advances in vehicles such as ships, airplanes, and tanks, especially in their engine designs and fuel enhancements. Improvements were also made to rockets, guided weapons, and aiming devices (which later helped with our space program). Scientific progress not only continued during the war, but in many ways was accelerated by it—a beneficial consequence, not a justification. Yet, these good things were forged under intense pressure to defeat profound, existential evil.

In 1928, near the beginning of his ascent to power, Adolf Hitler claimed, "We tolerate no one in our ranks who attacks the ideas of Christianity...in fact our movement is Christian."[12] However, Hitler was cloaking inherently unchristian and anti-

Christian policies in Christian terms. After the death of the German president in 1934, then-chancellor Hitler claimed both offices to become the Fuhrer—a dictator—and revealed his true beliefs by enacting those policies that systematically brought about atrocities on an unprecedented mass scale that could not in any way be justified biblically.

Hitler (CC-BY-SA 3.0)

Christianity's continued decline in Europe after World War I created a population still familiar with biblical terms, yet strangers to scripture. Hitler manipulated the ignorance to disguise his diabolical aims by baptizing them in ways accepted by a nominally Christian culture. Ironically, the most important theologian of the 20th century taught in Germany during Hitler's rise to power and recognized exactly what was happening.

Karl Barth (by Album – Alamy Stock Photo)

Although Karl Barth (1886–1968) was Swiss, he was educated mostly in Germany. After serving as a pastor in Switzerland, he took positions in theology at several German universities beginning in 1921. Barth published a commentary on the book of Romans in 1919 that reaffirmed the reality of sin and the fallen nature of humanity. His articulate defense of the historic Christian doctrine of original sin and his massive work *Church Dogmatics* were a decisive blow against the theological liberalism that began with Schleiermacher and contributed to the progressive idea of innate human goodness. However, the influence of German higher criticism of scripture was evident in other areas of his thought. He believed God had revealed himself, but that only Jesus was infallible, not scripture itself. He believed God ruled creation and was involved in it, but did not have absolute sovereignty or perform miracles. His theology that reaffirmed some important doctrines while redefining others became known as Neo-Orthodoxy.

Before the end of the war, Barth had diagnosed how the Nazi rise happened, writing, "Western civilization failed to confront National Socialism firmly because the realization of the Christian revelation among the civilized people of the West (not only among the Germans!) had become dim. Men did not see the inherent atheism of the Hitlerian system."[13] However, Barth's generalization was not so broad that he didn't recognize a small, believing remnant of the church that stood in opposition. "There was resistance to Hitler from the very first on the part of those who were on their way back to a conscious realization of the Christian presupposition of Western culture....The first serious protest against Hitlerism necessarily had to come, and did in fact come, from the ranks of the Protestant Churches."[14]

Barth himself was an early vocal critic of the Nazis. As Fuhrer, Hitler claimed to be the head of the German Church. Barth responded by founding the Confessing Church movement along with pastors Martin Niemoller, Dietrich Bonhoeffer, and others. The consortium of churches, pastors, and theologians asserted their independence from government control in The Barmen Declaration, which was written almost entirely by Barth in 1934.[15] The next year, after refusing to sign an oath of allegiance to Hitler, Barth was forced to resign from the University of Bonn and return to Switzerland, where his freedom to speak out combined with his increasing fame amplified his voice worldwide.

Martin Niemoller was sent to the concentration camps and survived to be liberated. Bonhoeffer became a double agent for the Germans with the intent to work against them. He was even involved to some degree in at least one assassination attempt on Hitler. When the Nazis discovered he was using his position as a German agent to secretly facilitate escape routes for Jews, he was sent to a concentration camp as well. He was executed by hanging twenty-nine days before Germany surrendered.

In the German occupied countries of Europe such as France and The Netherlands, the church played a special role in thwarting the Nazis. It provided a conduit to help Jewish people escape by providing passports and transportation to the West. It also worked closely with the local underground resistance movements to pass military intelligence to the allies and thereby undermine German activities.

In England, the Church of England claimed the role of moral conscience of the country. It strongly supported the war although there was some disagreement about the morality of the mass bombing of German cities, which damaged some of the historic cathedrals. Despite the strong position of the church in society, the church membership and attendance actually declined during the war, as in other parts of Europe. An attempt was made to combine the Anglican and Methodist churches, but the details couldn't be worked out, and the attempt failed.

Although Protestants comprised two-thirds of Christians in Germany, the different denominations did not agree on how to deal with the Nazis. The twenty million Roman Catholics, however, spoke with one voice through Pope Pius XI.[16] Initially, Pius XI perceived Communism to be the greater threat and aggressively consecrated new bishops in the Soviet Union to counteract persecution of the church. But the USSR defeated the plan by sending the newly minted bishops to labor camps. The Pope used a different approach with the Nazis: he signed a concordat intended to protect the church. The agreement had the unintended consequence of legitimizing

Hitler's standing, increasing his power. By 1937, Hitler had violated the concordat so many times that Pius XI issued an encyclical to be read from the pulpits of German churches that condemned the Nazis. The concepts echoed Barth's critique of Western society and The Barmen Declaration's rejection of a national religion, racism, and anti-Christian doctrines. For example, "None but superficial minds could stumble into concepts of a national God, of a national religion; or attempt to lock within the frontiers of a single people, within the narrow limits of a single race, God, the Creator of the universe, King and Legislator of all nations before whose immensity they are 'as a drop of a bucket' (Bible: Isaiah 11:15)."[17]

Pope Pius XI

After Pope Pius XI died in 1939, Pope Pius XII took yet a different tack: impartiality. Although issuing condemnation of Nazi policies and offering Vatican City as a sanctuary for Jews and other refugees, Pius XII essentially positioned himself to be able to act as a mediator in the aftermath of the war. In 1998, Pope John Paul II issued an apology to the Jews for inaction by the church during the Holocaust.[18]

The threat of Communism to Christians in the Soviet Union—especially under Stalin—tragically vindicated Pope Pius XI's concern as it devasted the Orthodox Church. In 1914, at the end of Tsarist Russia, there were more than 54,000 churches. By 1939, the number was in the 100s. During that time, more than 50,000 priests and deacons were imprisoned, sent to concentration camps, or executed, reducing their number to less than 200. Over 1,000 monasteries, 57 seminaries, and four theological academies were closed.[19]

Early on, there was a major anti-war movement in the United States led by churches wanting to maintain peace. Unlike Europe, Christianity in the United States between 1926 and 1950 grew at more than double the rate of the population, significantly strengthening its influence.[20] Prominent citizens such as aviator Charles Lindbergh and businessman/ politician Joseph Kennedy supported the movement. As the war raged in Europe, the United States maintained its neutrality while providing a minimal amount of aid to England. However, when Japan attacked Pearl Harbor, the United States had no choice but to enter the war in both the Pacific and Europe. Conscientious objectors were allowed to serve as non-combatant medics or chaplains. In addition, the Civilian Public Service program was formed to give people the opportunity to do critical jobs related to the war effort without serving in the military. The church supported these activities and ministered to families experiencing the loss of loved ones.

The uniquely Japanese religion of Shintoism made them a completely different kind of enemy from the Nazis and informed how they fought. This was not the Asian version of the Nazi's ersatz national religion. In Shintoism, the island of Japan was created for the gods. The emperor himself was *kami*, one of the gods. Therefore, the Japanese people believed they had a divine origin that made them distinct from all other people, that whatever did not benefit Japan was sin, and that the purpose of life was to rejoin the ancestors in the afterlife. According to them, because it was an honor society that worshipped their nationality, war was inherently a holy war, with soldiers willingly sacrificing their lives. This radical devotion was borne out through the use of Kamikaze suicide bombers against allied ships in the South Pacific. The majority of Japanese were also Buddhists. Buddhism and Shintoism are not mutually exclusive, and many practiced both at the same time. Some Japanese Buddhists struggled to support the war because of their commitment to non-violence, but the principle of "compassionate killing" justified their involvement.

A kamikaze attack on the USS Essex in 1944

During the war, the Japanese government forced the small minority of Christians into a single denomination called the United Church of Christ in Japan. Secret police suppressed and persecuted the group because it did not acknowledge the Imperial state as the ultimate authority, making it a threat to Japanese culture.

The atomic cloud rising above Hiroshima in 1945

Japan trained their civilian population—including women and children—to fight for their country. If the allies had invaded the Japanese homeland to end the war it would have resulted in massive numbers of casualties on both sides. It was fortunate that the allies were able to use atomic weapons to compel the Japanese to surrender before invasion was necessary. As devastating as the bombs dropped on Hiroshima and Nagasaki were, the resulting casualties were far less than the estimated number of casualties during a ground invasion. After the end of the war, the Occupation imposed a dismantling of Shintoism as the state religion and the emperor renounced his claim a *kami*.

Shintoism had implications in the United States as well. Because of the national worship of Shintoism, many Japanese Americans near the West Coast were placed in internment camps after the attack on Peral Harbor. Since Buddhism was often conflated with Shintoism, a suspicion of Buddhism in America lasted into the 1960s.

Despite the rapid growth of science and technology during World War II, there was no significant conflict between science and religion. The far greater conflict with religion came from different governments. Werner Heisenberg was no doubt aware of what happened to Christians in Germany who openly opposed the Nazis, and it may have influenced him to leverage his position for clandestine resistance. His belief that science and religion are complementary, where his scientific work was done within the ethical boundaries of faith, was a silent, yet pivotal factor in the war. If he had resisted in more obvious ways, he may have been sent to the concentration camps, forfeiting his position to a scientist more sympathetic to the Nazi quest for an atomic bomb.

Heisenberg lived out the charge Bonhoeffer gave his students as he taught in an underground seminary. "Jesus Christ lived in the midst of his enemies...On the Cross he was utterly alone, surrounded by evildoers and mockers. For this cause he had come, to bring peace to the enemies of God. So the Christian, too, belongs not in the seclusion of a cloistered life, but in the thick of foes."[21]

References and Notes

1. Heisenberg, W., *Encounters with Einstein: and Other Essays on People, Places, and Particles* (New York: Seabury Press, 1983), 9.
2. Ball, P., How 2 Pro-Nazi Nobelists Attacked Einstein's 'Jewish Science,' https://www.scientificamerican.com/article/how-2-pro-nazi-nobelists-attacked-einstein-s-jewish-science-excerpt1/. (accessed 12.30.24).
3. Roberts, A., *The Storm of War* (New York: Harper Collins, 2011), 573.
4. Jungk, R., *Brighter Than a Thousand Suns* (New York: Grove Press, 1956, 1958), 103.
5. Margenau, H., "Why I Am a Christian," *Truth Journal*, vol. 1, 1985, https://www.leaderu.com/truth/1truth16.html (accessed 1.1.25).
6. Heisenberg, W., *Encounters with Einstein: and Other Essays on People, Places, and Particles* (New York: Seabury Press, 1983), 9.
7. Heisenberg, W., *Physics and Beyond* (New York: Harper and Row, 1972), 83.
8. Ibid, 83–84.
9. Heisenberg, W., *Across the Frontiers* (New York: Harper & Row, 1974), 213.
10. Ibid, 19.
11. *Book of Concord, The*, ch VI, https://bookofconcord.org/solid-declaration/third-use-of-the-law/ (accessed 01.03.25).
12. Steigmann-Gall, Richard, *The Holy Reich: Nazi Conceptions of Christianity, 1919–1945* (New York: Cambridge University Press, 2003), 60.
13. Barth, Karl. "The Protestant Churches in Europe," *Foreign Affairs* 21, no. 2, 1943: 265. https://doi.org/10.2307/20029223.
14. Ibid.
15. The Barmen Declaration can be read here: https://creedsand-confessions.org/barmen-declaration.html (accessed 01.03.25).
16. The German Churches and the Nazi State. https://encyclopedia.ushmm.org/content/en/article/the-german-churches-and-the-nazi-state (accessed 01.03.25).

17. *Mit Brennender Sorge,* Encyclical of Pope Pius XI, March, 13, 1937. https://www.vatican.va/content/pius-xi/en/encyclicals/documents/hf_p-xi_enc_14031937_mit-brennender-sorge.html (accessed 01.03.25).

18. Pope John Paul II, We Remember: A Reflection on the Shoah, http://www.christianunity.va/content/unitacristiani/en/commissione-per-i-rapporti-religiosi-con-l-ebraismo/commissione-per-i-rapporti-religiosi-con-l-ebraismo-crre/documenti-della-commissione/en1.html (accessed 01.03.25).

19. Ware, T., *The Orthodox Church* (New York: Pelican 1963; revised 1993), 148, 162.

20. Herberg, W., *Protestant, Catholic, Jew: An Essay in American Religious Society* (New York: Doubleday, 1955), 60.

21. Bonhoeffer, D., *Life Together* (New York: Harper & Roe, 1954), 17.

Chapter 7

Charles Townes (1915–2015) and Arthur Schawlow (1921–1999): A Solution in Search of a Problem

The heavens declare the glory of God;
the sky displays his handiwork.
Day after day it speaks out;
night after night it reveals his greatness.
There is no actual speech or word,
nor is its voice literally heard.
Yet its voice echoes throughout the earth;
its words carry to the distant horizon.
In the sky he has pitched a tent for the sun.

Psalm 19:1–4

Seven Saints of Science: An Introduction to the History of Conflict between Science and Religion
Richard C. Powell and Doug Powell

ISBN 978-981-5352-24-5 (Hardcover), 978-1-042-00201-6 (eBook)
www.jennystanford.com

DOI: 10.1201/9781042002016-7

"I see religion as an attempt to understand the purpose of our universe and science as an attempt to understand its nature and characteristics, so that the two are necessarily closely related."[1]

Charles Townes

"The context of religion is a great background for doing science... Thus scientific research is a worshipful act, in that it reveals more of the wonders of God's creation."[2]

Arthur Schawlow

Townes and Schawlow (© Doug Powell)

7.1 Townes' and Schawlow's Science and Faith

What would you say are the most important inventions of the 20th century? Most people would put the transistor and the laser as the top two on their list. The first of these stimulated the growth of microelectronics, and the second has found applications in every aspect of our daily lives. These inventions took place shortly after World War II which was an era of high-tech development. The important war-time development of technologies such as radar made the public appreciate the value of research and development. After the war, society continued to invest heavily in basic and applied research, and the invention of the laser was a direct result of this support.

The two physicists generally given credit for the inventing the laser are Charles Townes and Arthur Schawlow. In 1964, Charles Townes was awarded the Nobel Prize for the invention of the laser which he shared with Russian scientists N.G. Basov and A.M. Prokhorov, who developed gas masers (a microwave predecessor of the laser) and semiconductor lasers. In 1981, Arthur Schawlow shared the Nobel Prize with two other scientists for the interaction of light with matter using lasers. More than thirteen Nobel Prizes in physics as well as several others in chemistry have been associated with the use of lasers. In an early interview with Arthur Schawlow, he was asked *What is the laser good for?* He answered, "It is a solution in search of a problem." Lasers are now used to perform thousands of tasks in every aspect of our daily lives. Interestingly, despite the great variety of applications and the innumerable problems it solves, the laser came about through curiosity-driven research rather than research directed to the solution of a specific problem.

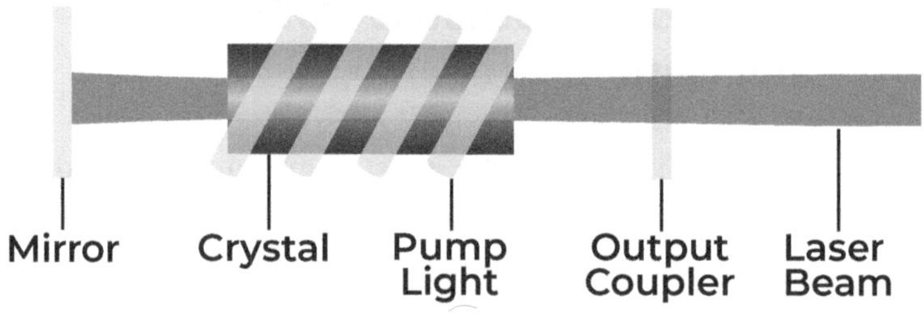

Laser Diagram (© Doug Powell)

Even though the laser has become familiar in our lives, most people can't define exactly what it is. The word *laser* is an acronym for Light Amplification by Stimulated Emission of Radiation. Thus, lasers are based on the process of an atom emitting light by the process of "stimulated emission." Remember that when an atom absorbs or emits light, it means that one of the electrons on the atom moved from one of its orbits to another one increasing or decreasing its energy. Usually, when an electron is raised to a high-energy orbit, it only stays in the upper energy state just for a brief amount of time and then decays to a low-energy state, spontaneously emitting light. In the early years of the 20th century, Albert Einstein postulated that if an electron stayed in the excited state long enough, instead of decaying spontaneously, it could be stimulated to decay to a lower state. This would occur if the electron was hit by a photon of light having the same energy as the energy difference between the two electronic states. When this occurred, the initial photon and the stimulated photon would leave the atom in unison going in the same direction, with the same energy. The light beam emitted at a specific frequency would be stronger than the one that hit the atom because it would have more photons.

When spontaneous emission occurs from a substance, all of its atoms act independently. This means the light emission goes in all directions with different colors and different wave patterns. The beam from stimulated emission is called "coherent light" because the photons act together instead of acting independently. This means that the light waves produced by stimulated emission from a group of atoms all have the same color and wave patterns and are going in the same direction. A coherent beam of light can be focused to a smaller point, emitted as a shorter pulse, amplified to a higher power, and transmitted over longer distances compared to light produced by spontaneous emission. The many applications of lasers depend on at least one of these special properties.

Some lasers are as small as the eye of a needle, and some are as large as a big building, but they all operate the same way. The design of a specific laser depends on the properties needed for its intended application. All lasers are made up of three parts. The first is a substance containing atoms that absorb energy and emit light with an excited electron orbital that has a long enough lifetime for stimulated emission to occur. The substance can be a solid, a liquid or a gas. The second part of every laser is a source of input energy. For crystalline or glass solids this can be a normal flash lamp. For gases, this can be an electrical discharge like a fluorescent light. For semiconductors, this can be an external electrical current. Chemical reactions can also provide the input energy input for certain cases. The third element of every laser is an optical cavity. This is made of two mirrors. One is 100% reflecting, and the other is a partial reflector that allows some of the light hitting it to get out. When the input energy excites the atoms of the active substance, they spontaneously emit light, and as the light reflects back and forth through the optical cavity, stimulated emission occurs, and a laser beam comes out the partially reflecting end of the cavity.

The development of this ingenious invention is an interesting story. In 1958, brothers-in-law Arthur Schawlow and Charles Townes published a paper in a physics journal entitled "Infrared and Optical Masers."[3] This classic paper described exactly how to make a laser. However, the laser they built using ruby did not work. Two years later, Ted Maiman at Hughes Research Labs followed the same procedure to build a laser using ruby that *did* work. Later, it was learned that having the right concentration of optically active chromium ions in ruby crystals and having input energy over a specific threshold is critical to make this type of laser work.

Charles Townes

Townes and Schawlow met when Townes hired Schawlow to join his lab at Columbia University as a postdoctoral fellow. Although they had a highly productive professional collaboration, Schawlow said the best thing that came out of their interaction was Townes introducing him to his sister, Aurelia, whom he married. Unfortunately, since Townes was becoming the department head, the marriage ultimately meant

Schawlow had to leave Columbia to escape the appearance of nepotism, and he took a position at Bell Telephone Laboratories in New Jersey. Fortunately, Bell Labs was only an hour away from Columbia, allowing them to continue to collaborate on weekends. Townes was consulting for Bell Labs at the time they published their landmark paper on the invention of the laser, giving Bell Labs the original patent. The brothers-in-law each went on to have distinguished careers in science and public service.

Charles Townes is recognized as one of the most important experimental physicists of the 20th century. In addition, he served as an advisor to every presidential administration from Truman in 1945 to Clinton in 1999. Townes was born in 1915 in Greenville, South Carolina. He graduated from Furman University as a double major, earning bachelor's degrees in physics and modern languages. During his time there, he was on the swim team, the sports band, and the college newspaper. He earned his master's degree in physics from Duke University, and a Ph.D from Cal Tech. His thesis research was in the area of nuclear physics.

Bell Labs recruited Townes for its technical staff during World War II. His research focused on the military use of radar, and he patented several new technologies in this area. In 1948, he became a professor at Columbia University, where he also served as Executive Director of the Columbia Radiation Laboratory, and eventually as Chair of the Physics Department. After several years as Vice President and Director of Research for the Institute for Defense Analyses in Washington, D.C., he became Provost and Institute Professor at MIT. In 1966, he accepted a position at the University of California, Berkeley, where he remained for the rest of his career.

The United States has developed into such a high-tech society that the government is heavily involved in both funding and oversight of the scientific community. In light of this, it is vital to have knowledgeable scientists as government advisors. As advisor to the government, Townes served as Vice Chairman of the Science Advisory Committee to the President, Chairman of the Advisory Committee for the first human landing on the moon, and Chairman of the Defense Department's committee on the MX missile. These important positions gave him significant influence on the scientific community in the United States. In addition, he served on the Board of Directors of General Motors Corporation and Perkin-Elmer Corporation.

Charles Townes with a maser

Townes' own research interests evolved significantly over his lifetime. His initial research was on microwave spectroscopy to study the structure of atoms and molecules. This led to his development and use of the maser (Microwave Amplification by Stimulated Emission of Radiation), which he patented. Working with Schawlow, they published the definitive paper describing how to move these concepts into the visible region of the spectrum to create a laser. He continued his research on lasers and nonlinear optics, but when he moved to Berkeley, he focused more on astrophysics. Townes made major contributions in radio and infrared astronomy. His research group detected the first complex molecules in interstellar space, and they measured the mass of the black hole in the center of our galaxy. Along with the Nobel Prize, Townes was awarded many other research prizes and given twenty-seven honorary degrees from universities around the world.

Although he was raised Baptist, Townes, his wife, and four daughters were active in churches of different denominations depending on where they lived at the time. He served as a deacon in an Episcopal church and was also a member of Congregational, Presbyterian, and Lutheran congregations. He began each morning reading scripture and praying with his wife and ended each day in prayer with her as well.[4] When he passed away in 2015, UC Berkeley colleague and astrophysicist Reinhard Genzel said, "His strength was his curiosity and his unshakable optimism, based on his deep Christian spirituality."[5]

In addition to his scientific work, Townes wrote essays and spoke on the interaction of religion and science. At a conference at the Vatican in 2001, he said, "I do not agree with those who maintain a high wall of separation between science and religion. Instead, I would rather admit that science and religion are not discretely separated domains. Rather, they constitute two complementary modes of human understanding; and at least in some areas they share a common domain of knowing. In fact, I hesitate to put boundaries on either one, and expect that as we understand each more fully, they will more fully overlap."[6] On the question of cosmogony, he said,

"I do not understand how the scientific approach alone, as separated from a religious approach, can explain an origin of all things...The question of origin seems always left unanswered if we explore from a scientific view alone. Thus, I believe there is a need for some religious or metaphysical explanation if we are to have one."[6] In short, "Science tries to understand what our universe is like and how it works, including us humans. Religion is aimed at understanding the purpose and meaning of our universe, including our own lives. If the universe has a purpose or meaning, this must be reflected in its structure and functioning, and hence in science."[8] In 2005, Townes was awarded the Templeton Prize for his writings on science and religion. He gave half of the $1.5 million dollars in prize money to Furman University, and half to various church charities.

Arthur Schawlow (by Stanford University CC-BY-A 3.0)

Arthur Schawlow, was born in Mount Vernon, New York in 1921 to a Jewish immigrant father from Riga and a Canadian Protestant mother. He was raised as a Protestant in Toronto, Canada, and graduated from the University of Toronto at the start of World War II. During the war he taught at the University of Toronto and worked in a microwave factory developing radar. After completing his Ph.D. at the University of Toronto, he accepted a postdoctoral position at Columbia University with Charles Townes, the leading expert on microwave spectroscopy. At the end of his two-year postdoc, he became a research scientist at Bell Labs—the most prestigious research laboratory in the world at that time. Ten years later, in 1961, he took a position at Stanford University where he was a professor until his death in 1999.

His collaboration with Townes continued after Schawlow left Columbia, and they wrote the classic textbook, *Microwave Spectroscopy.*[9] At Bell Labs, his research focused on superconductivity and nuclear quadrupole resonance. Schawlow and Townes also worked on ways to extend the maser into the visible spectral region. In 1958, they published their seminal paper on how to make an optical maser, or the laser. From then on, Schawlow's work centered on using the special properties of lasers to develop new types of laser spectroscopy techniques which led to his Nobel Prize. One of the most important techniques he worked on was laser cooling of atoms which has become an important technique in atomic spectroscopy. He also worked on making lasers out of exotic substances such as Jell-o, which led to a new type of laser configuration. In 1991, the American Physical Society honored him by establishing the Arthur L. Schawlow Prize in Laser Science, awarded annually to individuals doing outstanding research using lasers and funded by the NEC Corporation.

Schawlow was a Jazz enthusiast and played clarinet. His wife studied music formally and was the choir director at the United Church of Christ they attended with their two daughters. However, most of his time outside the lab was devoted to caring for their autistic son, who had to be institutionalized. Their advocacy for families with autistic members led to the founding of a new institution in Paradise, California, now called the Arthur Schawlow Center. Schawlow used the Nobel Prize money to buy a communication device that enabled his son to communicate after more than a decade of silence.

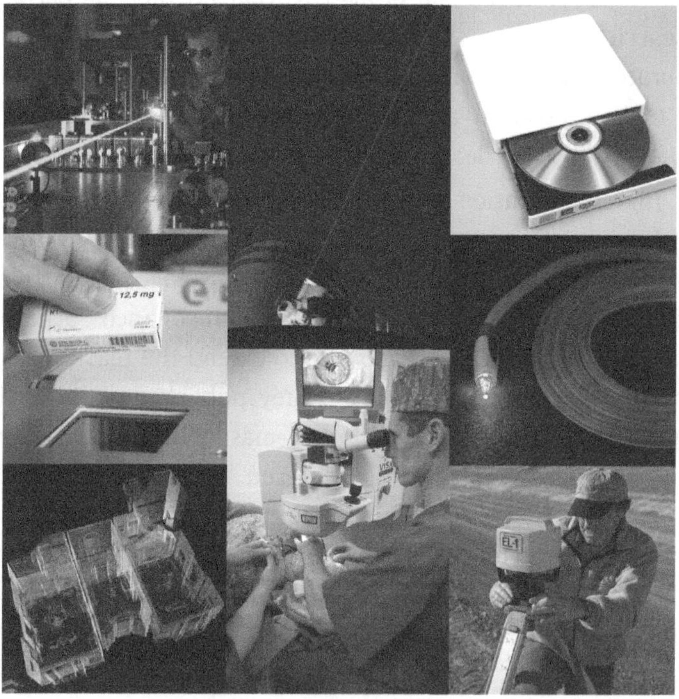

Examples of applications of lasers

Like his brother-in-law, Charles Townes, Schawlow's Christian faith framed all areas of his life. "The context of religion is a great background for doing science...Thus scientific research is a worshipful act, in that it reveals more of the wonders of God's creation."[10] And like Townes, Schawlow belonged to a several denominations during his life, including attending a Methodist church when he was visiting his son out of town. Although he did not write about the interaction of science and religion, he often shared his thoughts about it in interviews. For him, science and religion answered two different questions asked about the same thing. "When confronted with the marvels of life and the universe, one must ask why and not just how. The only possible answers are religious. For me that means Protestant Christianity."[12] Schawlow considered Einstein the greatest scientist of all time, Townes the greatest he had personally met, and Faraday the one he most admired.[13]

Charles Townes and Arthur Schawlow were excellent examples of world-class scientists who were also men of strong faith. Their commitment to make the world a better place extended beyond their labs and into society, both locally and globally.

7.2 The Church in the High-Tech Era

In some ways, the future prophesied by early 20th-century progressives such as H.G. Wells arrived after World War II like a familiar stranger materializing out of a fog of hope and promises. Penicillin put bacteria in check, solid-state transistors miniaturized electronics, and Chuck Yeager flew faster than the speed of sound—all before the 1940s had come to a close. The high-tech era gained momentum in the 1950s with advances such as the discovery of DNA, the first nuclear power plants and commercial computers, the inventions of the solar panel and the laser, Sputnik 1 orbiting the Earth, and color television broadcasting coast-to-coast (albeit onto mostly black-and-white sets). The 1960s saw developments literally out of this world, beginning with Uri Gagarin becoming the first man in space in 1961, and ending with Neil Armstrong walking on the moon in 1969. Between these bookends were the first organ transplants, CT scans, industrial robots, and a protocol for networking computers that would grow into the internet.

Buzz Aldin walking on the moon in 1969

But as always, the new heights cast longer shadows. In 1952, the first test of Edward Teller's Hydrogen bomb proved 700 times more powerful than the atom bomb dropped on Hiroshima. In 1962, the USSR detonated the largest thermonuclear bomb ever tested, which was still five times more powerful. The arms race created an industry that fueled the Cold War and produced thousands of nuclear warheads. For the first time in history, the human race had the technical ability to destroy itself and the political tension to make it a real possibility. The proxy wars of Korea and Vietnam lacked the moral clarity of World Wars I and II, engendering cynicism and mistrust. In 1947, scientists concerned with the abuse of technology created the metaphor of the Doomsday Clock to signify how close the Earth was to human-made global catastrophe, a metric still in use.[13]

The high-tech era birthed by America's Greatest Generation[14] arrived at the pinnacle of religious adherence in the history of the United States.[15] In 1800, less than 10% of the population were members of a church.[16] In 1906, 39% belonged to a religious body.[17] By 1956, the number had risen to 60%—an increase double the rate of population growth[18]—with 95% self-identifying as Christian or Jewish.[19] The unprecedented religious majority's Judeo-Christian framework gave a measure of restraint and offered meaning to the science reshaping society.

The modernist-fundamentalist controversies that built up to the Scopes Monkey trial in 1925 exposed a fissure in American Protestantism that continues to this day. So-called mainline churches (Congregationalist, Episcopalian, Lutheran, Methodist, Presbyterian, Northern Baptist, and Disciples of Christ)[20] by-and-large, embraced progressive, liberal theology. The Fundamentalist churches remained committed to the historic Christian understanding of the inspiration of scripture, divinity of Christ, Original Sin, and the penal substitution view of the atonement. Each of the mainline denominations has branches that split off to preserve the same conservative doctrines.

While mainline Protestants and Roman Catholics remained politically engaged and were deeply involved in social issues as well as in the arts and Hollywood, after the Scopes Trial's marginalization of Fundamentalists, Evangelicals basically withdrew from the political sphere to focus on evangelism. Missionaries were sent from America to Europe after discovering their majestic cathedrals and historic churches stood as empty shells after the war. In the United States, advances in technology gave evangelists new tools to proclaim the Gospel on a mass scale. Several influential preachers were early adopters of radio, creating regional and national pulpits, setting the stage for later televangelists. The televised crusades of Billy Graham extended the revivalist tent meetings of the Second Great Awakening into millions of homes around the world over six decades, preaching the Gospel to more people than any other person in history.[21] And as his crusades became more popular, he used them to advance the cause of civil rights.

Billy Graham preaching in Norway in 1955

Graham not only forced his audience to confront the issue by integrating the crusades; he gave a platform for Martin Luther King to expand his audience. King took the fight beyond the moral sphere and into the political sphere, becoming an icon of Christian engagement in society. Interestingly, King was highly critical of the Fundamentalism preached by Graham, calling it "contrary to science."[22]

Martin Luther King, Jr. in 1963

As the children of the Greatest Generation came of age, they questioned the faith of their parents, notwithstanding the popularity of Graham and King. The Boomers may have been named for the explosion of babies after World War II, but it also described the turbulence they brought to the 1960s as they came of age. Protest to the Vietnam War grew throughout the decade as trust in the government eroded, especially in the wake of the assassinations of the Kennedy brothers and King. Cynicism toward institutions and conventions led some to explore Eastern traditions and philosophies, expanding the New Age movement in the West which up until then had held only niche appeal. Others abandoned religion altogether, declaring, "God is dead."

The Roman Catholic Church responded by calling the first church council not convened to address heresy, but to renew and modernize the church. Vatican II (1962–1965) replaced the Latin liturgy with modern vernacular. The mutual anathemas between Rome and the Eastern church issued in 1054 were revoked, undoing (on paper at least) the Great Schism. A commitment to human rights, the poor, ecumenism, and religious tolerance significantly changed how the Catholic church engaged culture. And yet the number of men entering Catholic seminaries dropped by a third during the 1960s.[23]

Vatican II council meeting

The jettisoning of authority and tradition had profound implications for sexual ethics and the family. The sexual revolution brought with it unintended pregnancies. Pope Paul VI's proscription against birth control corroborated the irrelevance of the church to those who had left it. Along with the movement for equal rights of women, the desire for sexual freedom included freedom from consequences, leading to the legalization of abortion throughout Europe and the US.

Roe v. Wade, the US Supreme Court's 1973 ruling protecting the right to abortion, awoke the Evangelical political conscience that had been mostly dormant since the Scopes Trial 50 years earlier. The so-called "Religious Right" or "moral majority" grew more organized throughout the 1970s, showing its power by contributing significantly to the election of Ronald Reagan in 1980. In addition to abortion, traditional views of family, marriage, sexuality, and morality were fundamental tenets that united Christians from across a broad swath of denominations. Evangelicals have remained politically engaged, although their percentage has decreased steadily since then.

Mainline denominations who focused more on social issues frequently found themselves at odds with Evangelicals. And yet both have experienced decline since the 1950s. The 1960s began with 93% of Americans self-identifying as Christian. By 2000, the number had dropped to 77%.[24] At the same time, the people who have no religious conviction rose from less that 5% to more than 15%.[25] If Arthur Schawlow was right about science describing the "how" of the world, and religion explaining the "why," meaning is on a downward trend.

Charles Townes and Arthur Schawlow expressed their faiths in how they approached their work and how they lived their lives. Both were active in their churches, including their social programs. Townes believed, "scientific evidence leads to a feeling and hypothesis that the course of natural history is guided at least in part by an intelligence and a purpose."[26] The *how* pointed to a *why*. For them, science was like studying the physical properties of a coin, and religion was studying the images and text imprinted on the coin. They are inextricably linked.

References and Notes

1. Townes, C., "The Convergence of Science and Religion," ASA Annual Meeting, Pepperdine University, August 2, 2002.
2. Margenau, H., Varghese, R.A., eds., *Cosmos, Bios, Theos* (La Salle: Open Court, 1992), 106.
3. Schawlow, A.L., Townes, C.H., "Infrared and Optical Masers," *Physical Review*, vol. 112, 1940 (1958). DOI: https://doi.org/10.1103/PhysRev.112.1940.

4. Townes, C.H., *A Life in Physics: Bell Telephone Laboratories and World War II, Columbia University and the Laser, MIT and Government Service, California and Research in Astrophysics.* an oral history conducted in 1991–1992 by Suzanne B. Riess, Regional Oral History Office, The Bancroft Library, University of California, Berkeley, 1994, 605–606, 615–617.

5. UC Berkeley News, January 27, 2015. https://news.berkeley.edu/2015/01/27/nobel-laureate-and-laser-inventor-charles-townes-dies-at-99/ (accessed 01.07.25).

6. Townes, C., "Logic and Uncertainties in Science and Religion," *Science and the Future of Mankind*, Pontifical Academy of Sciences, Scripta Varia 99, Vatican City 2001.

7. Townes, C., "The Convergence of Science and Religion," ASA Annual Meeting, Pepperdine University, August 2, 2002, 122–123.

8. UC Berkeley News, January 27, 2015. https://news.berkeley.edu/2015/01/27/nobel-laureate-and-laser-inventor-charles-townes-dies-at-99/ (accessed 01.07.25).

9. Townes, C.H., Schawlow, A.L., *Microwave Spectroscopy* (New York: Dover, 1955).

10. Margenau, H., Varghese, R.A., eds., *Cosmos, Bios, Theos* (La Salle: Open Court, 1992), 106.

11. Margenau, H., Varghese, R.A., eds., *Cosmos, Bios, Theos* (La Salle: Open Court, 1992), 105.

12. Brian, Denis, *God Talk* (New York: Plenum Press, 1995), 253.

13. https://thebulletin.org/doomsday-clock/timeline/ (accessed 01.06.25).

14. The Greatest Generation is defined by Pew Research Center as born between 1901 and 1927. Generally speaking, they are the generation who served in WW II. https://www.pewresearch.org/politics/2016/12/15/americans-name-the-10-most-significant-historic-events-of-their-lifetimes/pp_12-15-16_historyevents-00-06-png/ (accessed 01.09.25).

15. Herberg, W., *Protestant, Catholic, Jew* (Garden City: Doubleday, 1955), 59.

16. Garrison, W.E., "Characteristics of American Organized Religion," *The ANNALS of the American Academy of Political and Social Science*,

vol. 256, no. 1, 1948, 14–24, https://doi.org/10.1177/000271624825600103.

17. Wolf, Richard C., "Religious Trends in the United States," *Christianity Today*, vol. III, no. 15, April 27, 1959, 3.
18. Ibid.
19. Herberg, W., *Protestant, Catholic, Jew* (Garden City: Doubleday, 1955), 59.
20. Hutchison, W.R., *Between the Times: The Travail of the Protestant Establishment in America, 1900–1960* (New York: Cambridge University Press, 1989), 4.
21. https://billygraham.org/about/biographies/billy-graham/ (accessed 01.09.25).
22. King, M.L., "The Sources of Fundamentalism and Liberalism Considered Historically and Psychologically," essay written at Crozer Theological Seminary, 1949, https://kinginstitute.stanford.edu/king-papers/documents/sources-fundamentalism-and-liberalism-considered-historically-and (accessed 01.11.25).
23. Shelley, B.L., *Church History in Plain Language* (Nashville: Nelson, 1982, 1995), 459.
24. Korhonen, V., Self-described religious identification of Americans 1948–2023, December 12, 2024. https://www.statista.com/statistics/245478/self-described-religious-identification-of-americans/ (accessed 01.11.25). This statistic does not reflect church membership or active attendance, merely self-identification.
25. Pew Research Center, September 8, 2022. https://www.pewresearch.org/religion/2022/09/13/how-u-s-religious-composition-has-changed-in-recent-decades/pf_2022-09-13_religious-projections_01-01-png/ (accessed 01.11.25).
26. Townes, C., "Logic and Uncertainties in Science and Religion, Science and the Future of Mankind," Pontifical Academy of Sciences, *Scripta Varia* 99, Vatican City 2001.

Chapter 8

Summary and Conclusions

Who has measured out the waters in the hollow of his hand, or carefully measured the sky, or carefully weighed the soil of the earth, or weighed the mountains in a balance, or the hills on scales?

Isaiah 40:12

8.1 Summary

There is a popular misconception among people not involved in science that religious belief among scientists is rare. Not only is this historically false, but it is not true now. A 2016 survey of biologists and physicists from around the world revealed almost 55% of scientists identify with a religion, including 39% in the United States and 37% in the United Kingdom. More than 20% have no doubt about the existence of God, including 10% in the United States and 8% in the United Kingdom. They also report a weekly attendance of religious services as 15% worldwide, 11% in the United States, and 8% in the

Seven Saints of Science: An Introduction to the History of Conflict between Science and Religion
Richard C. Powell and Doug Powell

ISBN 978-981-5352-24-5 (Hardcover), 978-1-042-00201-6 (eBook)
www.jennystanford.com

DOI: 10.1201/9781042002016-8

United Kingdom. For context, weekly church attendance in the United States is 33% of the population. Most interestingly, 68% of scientists believe there is no conflict between science and religion, including 63% in the United States and 59% in the United Kingdom.[1] That means even among non-religious scientists, there is a strong belief that science and religion do not conflict.

In this brief history of the conflict between science and religion, the focus has been on physics as the science and Christianity as the religion (except for Einstein). However, as the 2016 survey revealed, the history is similar for other fields of science and other religions. The choice of seven specific scientists was somewhat arbitrary since many other outstanding scientists who were people of strong faith existed during this time. The ones we have chosen should be considered as good examples rather than exceptions to the rule. Unfortunately, during much of this time period, women did not have the opportunity to fully participate in the scientific community. In the last 50 years or so, the landscape has changed significantly, and we could name female examples such as Donna Strickland, 2000 Nobel Prize winner in physics and an active member of the United Church of Canada.

Geneticist Francis Collins, a modern scientist with strong religious faith, recognized the fact that there actually is no conflict between science and religion. Instead, it is some people's interpretation of science or of religion that is in conflict. According to Collins, "It is one of the great tragedies of the last 150 years, in the United States, that an ultra-literal reading of the first chapters of the Book of Genesis has been taken as a litmus test for serious Christian faith."[2] Collins argued that the "powerful, mystical" words in Genesis "tell us about who we are, and who God is, but were never intended to be a scientific textbook.... The God of the Bible is also the God of the genome."[3]

Francis Collins

The profiles in this book span more than three centuries and describe the lives of some of the greatest scientists who ever lived and their strong religious faith. Before the scientific era, there was very little interest in how the world worked. Astrology and elements of Greek philosophy were the extent of science—hardly worthy of the name. But after Copernicus reoriented the night sky, the scientific revolution snowballed into a force that touches every area of life. Only religion has a more comprehensive influence because its explanatory power encompasses the metaphysical. The key is how to understand the relationship of these two authorities. Since opposing views of creation form a major area of conflict between science and religion, it is important to understand what is really meant by creation, intelligent design, and evolution.

In 1802, Christian apologist William Paley opened his influential book *Natural Theology* with a famous thought experiment. Imagine walking through a field and knocking your foot against a stone. If you wondered how the stone got there, you might answer that it had been there forever. As you walk a bit further, you discover a watch on the ground. If you wondered how the watch got there, it would be absurd to think it had been there forever. Why? Because the watch contains parts that work together for a purpose. In other words, it exhibits design, which infers a designer. The argument fits alongside Aristotle's classic argument for a prime mover or first cause.

Paley's watch (© Doug Powell)

Inference of design is an interpretation of data not everyone agrees on, just as there are different interpretations of scripture. As a college student, Charles Darwin himself was convinced by Paley's watchmaker argument, but recanted even before formulating the law of natural selection. In the intervening years, he had abandoned the faith.[4] Was Darwin guilty of confirmation bias—seeing what he wanted to see—or was he following the evidence? It is a question that must be asked not only of each scientist, but also of ourselves as we consider the evidence.

Areopagus (© Doug Powell)

People tend to accept either evolution or creation based on personal feelings rather than any evidence. Physicist and philosopher Blaise Pascal (1623–1662) wrote, "There is enough light for those who only desire to see, and enough obscurity for those who have a contrary disposition."[5] Or, as Christian apologist/philosopher Cornelius Van Til (1895–1987) put it, "The starting-point, the method, and the conclusion are always involved in one another."[6] It seems, then, that each of the four ways science and religion relate—conflict, independence, dialogue, and integration—says something about philosophical presuppositions as much as where the evidence leads. And because philosophical presuppositions are ultimately religious presuppositions to some degree, there must be some kind of relationship between science and religion. As astronomer Robert Jastrow (1925–2008) wrote, "For the scientist who has lived by his faith in the power of reason, the story ends like a bad dream. He has scaled the mountains of ignorance; he is about to conquer the highest peak; as he pulls himself over the final rock, he is greeted by a band of theologians who have been sitting there for centuries."[7]

When the Apostle Paul stood on Mars Hill in Athens and preached, he gave philosophical arguments for God's existence and used cultural references to make his case (Acts 17:16–34). At other times, he argued for the faith based on the evidence for the resurrection of Jesus, including listing eyewitnesses who were still alive and could be questioned (1 Corinthians 15:3–8). And when he wrote to the church in Rome, he argued that evidence for God is plainly seen in creation (Romans 1:18–20). Science was not a category of knowledge in the first century, and so appealing to it as evidence for God never occurred to early Christians. In fact, the most powerful apologetic was not in Christianity's arguments, but in its character and ethics. The way Christians cared for each other and treated those around them is what attracted people to the faith.[8]

When the scientific era began, it was primarily understood in a religious context, which was the prevailing worldview.

Investigating the world revealed something about God. Science grew in influence to become one of the two major pillars of society along with religion. Advances in science sometimes helped purge the church of mistaken teachings, such as with Galileo. But the conflict proved to be an agent that refined theology, rather than contradict it. The contemporary portrayal of science and religion as adversaries betrays an ignorance of both.

Two Pillars (© Doug Powell)

The scientists featured in this book embraced both pillars congruently. They all felt that their scientific research was a religious activity, though they each had different perspectives on religion. Lord Rayleigh, Werner Heisenberg, Art Schawlow, and Charles Townes were all members of traditional Christian denominations. Isaac Newton had biblically based beliefs that differed from Christian orthodoxy. Michael Faraday belonged to a small, forgotten church sect that was orthodox but rejected the state involvement in religion. And Albert Einstein invented his own religion (not as a cult leader, but as its sole adherent). Newton, Einstein, and Townes spoke and wrote prolifically about religious beliefs. Heisenberg and Schawlow wrote very little about their faith but openly discussed it. Rayleigh and Faraday expressed their faith by example, through involvement with their churches, and a commitment to reading scripture.

The lives, work, and public service of each of these men were motivated by their religious beliefs. Newton and Rayleigh served in politics, and they all served as advisors to different government committees. Rayleigh, Heisenberg, Townes, and Schawlow all contributed significant support to charities. Faraday was committed to being a good steward of the environment, and Einstein was an outspoken pacifist.

During the lifetimes of these scientists, the church underwent significant changes as it wrestled with how to engage this new field of expanding knowledge. Were scientists prophets of the material world reading a previously unknown language? Were they like rebel angels, about to be cast out of heaven for the prideful sin of believing they didn't need God? Or were scientists like the visible church on the Earth—a mixture of both that would still be used in God's providence for good? Ultimately, it is irrelevant as far as the truth is concerned since the church's doctrine of common grace welcomes all truth regardless of who discovers it.

8.2 Conclusions

In the 21st century, physics has become divided into big and small science. Big science requires enormous facilities such as the particle accelerator at CERN in Switzerland used for elementary particle research or the National Ignition Facility at Lawrence Livermore National Laboratory in California used for research on laser-induced fusion. Small science deals with microelectronics such as the design of computer chips to carry out specific tasks. These two types of science have little in common with each other and have resulted in very different cultures. In a way, this bifurcation is an allegory of the divisions we see in society. Politics, social issues, and religion have become increasingly polarized, creating a fractured culture.

CERN (by Maximilien Brice _CERN CC-BY-SA 3.0)

Accelerators on a chip (U.S. Department of Energy)

But there is more to the allegory. For all their differences, big science and small science are subject to the same laws of nature. They are united by common tenets that govern the work of both. They may focus on different parts of the world, but it is the same world. The late Senator Joe Lieberman recognized a similar connection in society that he believed could build bridges between societal factions, if not heal them. In his book, *Faith's Answers to America's Political Crisis,* Lieberman claims more than 90% of the people he represented in politics believed in God.[9] He contends that no matter what religion they are part of, living out their faith is an area of moral commonality that can transcend division for peaceful co-existence. Similarly, Christian apologist C.S. Lewis compiled examples of the moral code from world religions for the appendix of *The Abolition of Man* to demonstrate their strong similarities.[10] Stretching the scientific allegory to its limits, the hydrogen bomb's greater force than the atomic bomb proved unity (fusion) is more powerful than division (fission).

In the tension between science and religion, it turns out that the core issue is no different than in any other cultural arena, be it politics, entertainment, business, or academia: People of faith have a different worldview than atheists and agnostics. The great majority of the time, division is not over facts and data, but interpretation. Although there is a temptation for people of faith to appeal to God as an explanation to fill in the gaps of what we do not know scientifically, it is a lazy, unscientific attitude that only stifles scientific inquiry, and in fact, would have prevented science from progressing to its current advanced state. Conversely, there is a temptation for atheists and agnostics to make metaphysical—even religious—claims baptized in scientific language, which is equally unscientific and, in fact, outside the purview of science. Presuppositions delineate what can be admitted into the pool of possible explanations.

The scientists profiled in this book understood that both nature and scripture are necessary for understanding the world and our place in it. They each held deep religious convictions while conducting their scientific inquiry using methodological naturalism—the same approach as atheists. They looked for natural explanations to natural processes. They didn't do "Christian" science; they did science and they were Christians (excepting Einstein). The excellence of their work was recognized and accepted by the scientific community regardless of religious commitments. Their convictions that science reveals and supports the God they have faith in did not prevent them from doing world-class science. Including religious beliefs in presuppositions about the world does not mean scientific results are intrinsically flawed. These seven scientists demonstrate that.

Illustration of the Big Bang (by NASA - WMAP Science Team)

So what is the scientific view of the creation of the universe? Scientific evidence indicates that space and time began around 13.5 billion years ago with the Big Bang. All the matter in the universe today instantaneously appeared at a singular point that exploded into elementary particles. The laws of nature appeared at the same time, governing an evolutionary process that organized the elementary particles into atoms, then molecules, stars, and galaxies. When stars began to die, the laws of physics worked to produce additional elements that were distributed throughout the rest of the universe. The evolution of the universe is still underway, expanding and changing. Thus, science leaves the most basic questions of creation unanswered—who, how, and why.

Evidence such as the expansion of the universe and the existence of cosmic microwave background radiation leads back to a first moment in time where all matter was concentrated in a singularity, like a period punctuating the end of what can be observed. It is the delineation between existence and nonexistence of the physical universe. It is literally the end—or beginning—of science. But the period is also part of a question mark. Where did the matter come from? Where did the laws of nature come from? How did something come from literally nothing? Ironically, atheists who reject the miracle of the virgin birth of Jesus are confronted with the virgin birth of the universe.

The Big Bang is an effect, and every effect has a cause. But if nothing exists outside of the universe as the cause, the universe could not exist. The inference to a transcendent, uncaused cause was one of the reasons astronomer Fred Hoyle, an atheist, rejected the theory that the universe had a beginning and coined the pejorative name "Big Bang" in a 1949 BBC radio program. On another radio show, he claimed, "The reason why scientists like the 'Big Bang' is because they are overshadowed by the book of Genesis. It is deep within the psyche of most scientists to believe in the first page of Genesis."[11] But Hoyle's criticism of Big Bang advocates as tainted by presupposition was also true of himself given that he was an atheist who rejected the evidence for nonscientific reasons. Regardless of presuppositions, the question remains: What are the necessary preconditions for the universe coming into existence?

1. An entity that must be transcendent from the universe. The universe, as an effect, needs a cause that exists apart from it in order to bring it into existence.
2. The entity must have unimaginable power in order to create the universe—not merely the matter, but also the laws that govern it and the laws of logic that make it intelligible.
3. The entity must be self-existent. If the cause of the universe also had a cause, then we would look for the cause of the cause, beginning a chain of regression to the first, uncaused cause. The chain cannot be infinitely long because an infinite number of causes prior to creation makes it impossible to arrive at creation since an infinite number cannot be exhausted.
4. The entity must be personal. Because the universe is not self-caused, it is contingent on an agent outside of itself to bring it into existence. The Big Bang needs a banger. The banger therefore is an agent with the ability to choose and a will to act. Only persons have wills and choose.

These four logical inferences based on the scientific evidence from the Big Bang create only a general sketch of a metaphysical reality that is not only plausible but necessary. And yet the four necessary preconditions add up to a description most religions would recognize as God. Many Eastern religions, as well as Einstein's Cosmic Religion, reject the first and fourth inferences (though they do not explain how their rejection is possible) and still arrive at God using only the second and third inferences.

Religious symbols

The inferences, however, don't provide enough evidence to indicate *which* God is in view. But they are enough to defend the argument that science and religion are not in conflict. This is borne out in the survey of religious belief among scientists mentioned at the beginning of this chapter, where more than half of biologists and physicists from around the world identify as religious.[12] Whether or not they have worked through the logic syllogistically, they recognize it intuitively. Which takes us back to the definition of *faith*, defined in the introduction. *Pistis*, the Greek word in the New Testament translated as *faith*, means firm persuasion (which requires reasons) and conviction (which requires evidence). Considering the scientific evidence, there are excellent reasons to believe in God.

As for which God, it is up to each person to examine the religions to see if their doctrines accurately describe the world as it is and our experience in it without contradicting itself. Those are the marks of a worldview that is true and can be lived out consistently. Obviously, people will disagree in their conclusions, and even among adherents of a particular religion, believers will disagree on how to understand scientific evidence on particular issues. Flawed interpretations of scripture and/or of scientific data is the cause of the appearance of conflict between science and religion where none actually exists. Nowhere is this seen more clearly than in the question of the origin of life.

On the one hand are scientists who hold an atheistic/agnostic worldview. They not only argue the universe spontaneously came into existence from literally nothing; they argue life came from non-life. In both cases, the incredibly precise order in the universe and in living organisms is the product of random chance. By chance, the universe suddenly materialized into existence from nothing. By chance, physical laws came into existence at the same time, creating harmonious forces that ordered the matter into galaxies, solar systems, etc. By chance, the Earth was formed in exactly the right position to develop conditions that could sustain life. By chance, nonliving material became alive. By chance, the living organism evolved into more complex forms. To say the least, it is extremely difficult to see how science—a discipline of observation and experimentation that relies on the uniformity of the universe—can be based on random chance in any coherent way.

Nevertheless, they approach the data relevant to the origin of life through a lens of precision-by-chance. This is exemplified on the very first page of biologist Richard Dawkins' book, *The Blind Watchmaker*, where he states, "Biology is the study of complicated things that give the appearance of having been designed for a purpose."[13]

On the other hand are scientists who hold to a theistic worldview. Within the great variety of religious belief is a place for God as a causal explanation inferred by the evidence. They approach the data relevant to the origin of life through a lens of precision-by-design.

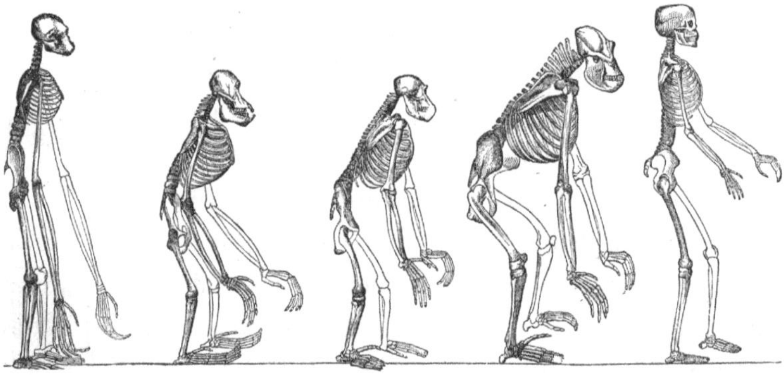

Illustration of the evolution of man from 1879

Since the most vocal conflict between science and religion is currently focused on evolution, it is important to address it directly. Straw men and misunderstandings on both sides complicate the issue unnecessarily, often making an emotionally explosive topic from what should be a consideration of the facts. Enumerating the facts to be interpreted is beyond the scope of this book, but a couple of things must be kept in mind as we work through the issue. Evolution through Natural Selection and random mutation as Darwin proposed has itself evolved. Contemporary evolution more often refers to Universal Common Descent brought about through changes within the DNA of organisms. Also, we need to distinguish macroevolution from microevolution. Macroevolution entails transition between species, while microevolution entails variation within species. Microevolution is universally accepted and not part of the conflict. This distinction is primarily emphasized by skeptics of macroevolution, while proponents of macroevolution often point to microevolution as part of the evidence they extrapolate from. In the conflict between science and religion, evolution refers to macroevolution.

There is a spectrum of interpretations of the data, and scientists holding religious beliefs can be found within each view except the first one.

- *Atheistic Evolution*: Evolution is well supported by the evidence and is a natural process that does not require God's involvement. Evolution accounts for both the origin of life and how life developed. Given enough time, random circumstances will become just right for the birth of life.
- *Theistic Evolution*: Evolution is well supported by the evidence and is a process directed by God or used by God as a secondary means. Evolution accounts for how life developed, but not its origin.
- *Scientific Agnosticism*: Evolution may prove to be true at some point but currently lacks the evidence to explain key elements in the process and has no evidence for life's origin. Universal Common Descent cannot be proven from the current evidence.
- *Creationism*: Evolution is not supported by the evidence. Life in all forms is the work of direct creation by God. Universal Common Descent is a misinterpretation of the facts.

Predictably, the extreme ends of the spectrum are where the loudest voices in the conflict lie. At one extreme are scientists who believe evolution is not just a process showing Universal Common Descent but also accounts for the origin of life itself. Whatever the evidence for evolution is, this overstates what can be explained by it. On the other extreme are scientists who view the evidence through the lens of a strictly literal interpretation of Genesis 1 and consider any other interpretation unbiblical. Clearly, these two perspectives cannot be reconciled. But it is each side's presuppositions fueling the conflict, not the evidence itself.

Between the poles, more nuanced views are held by scientists who want to be faithful to both the evidence and their religious beliefs, particularly in regards to the Bible. Conflict is not the only option. Some scientists see evolution as the process of life, while God is the agent who employs that process. They don't see evolution as a threat, but as an answer to the question of *how*, while God answers the questions of *who* and *why*. Theistic evolution is believed by many scientists and found a new expression in BioLogos, founded by Francis Collins, leader of the Human Genome Project. Theologians who contribute articles to BioLogos include N.T. Wright and Tim Keller, both prominent voices in the evangelical community. Within theistic evolution, there are different ideas of how evolution fits with biblical creation, but all incorporate it due to the evidence.

James Tour (by Jeff Fitlow CC-BY-SA 4)

Other scientists examine the same evidence without finding it compelling, yet do not reject the possibility that future evidence may sway the balance in its favor. For example, Chemist James M. Tour acknowledges the evidence for Universal Common Descent, but also that there is evidence within the far more prevalent "junk DNA" that are markers of uncommon ancestry. In addition, he points out evolution cannot demonstrate any real or proposed mechanism that enables transformations in body types, a prerequisite for Universal Common Descent. Until that can be shown, evolution should not be promoted as dogmatically unquestionable. As a Christian, Tour freely acknowledges that if evolution can be proven to be true, it would mean his interpretation of Genesis 1 is wrong, but not that the Bible is wrong.

Adding to the conversation is the concept of Intelligent Design. Although it is similar to the Natural Theology of Newton, Faraday, and Lord Rayleigh, who viewed their scientific discoveries not only as advances in knowledge of the world, but as proofs of God, Intelligent Design is intended to be more modest. ID is not an argument for God, let alone a specific god. Because design is not something the scientific tools and methods of chemistry, physics, biology, etc., are able to detect, ID is a philosophical argument. And because it is applied to scientific evidence, it is seen by some as pseudoscience. However, properly understood, ID's purpose is merely to detect information. Information is neither random, nor necessary, and therefore indicates design. ID theorist William Dembski has proposed a filter of criteria for recognizing design that includes the characteristics of contingency, complexity, and specificity.[14] If those conditions are met in objects in nature, then the presence of design infers a designer transcendent from this world.

Unfortunately, ID is often misunderstood and misapplied to prove the God of the Bible. Although ID is a bridge beyond the natural world, it does not entail a specific destination. That requires other arguments beyond the scope of ID. Additional arguments aren't necessarily a bad thing, but for apologists to conflate ID with other elements of a cumulative case creates confusion for what ID actually is.

Some atheistic evolutionists, such as Richard Dawkins, do not want religious influence to shape the articulation of science in the public sphere. But to question evolution does not mean there is a religious motive underlying it. And some Creationists refuse to acknowledge other possible interpretations of both the Bible and science. Intractable dogmatism on both ends of the spectrum is far more at the root of the evolution/creation conflict than science or scripture.

Many questions remain unanswered from these studies, and various geneticists and biologists have different interpretations of the data. What little is understood about prebiotic chemistry shows that the complex molecules and conditions required for life to begin could never occur randomly.[15] In addition, science has not identified the mechanisms that cause changes in complex molecules and does not know how they work. Until these mechanisms are understood, neither evolution nor any other scientific theory can explain the diversity and complexity of life on the Earth. As it stands, evolutionary theory cannot explain the existence or origin of life. Although scripture does have an explanation, it is not a scientific one. However, neither interpretation of scripture or of science should be considered immutable. Future knowledge in these areas is the key to finding agreement between scripture and scientific understanding. Truth is the goal of both science and religion.

In conclusion, science and religion are not incompatible silos of knowledge. There are different views for how they interact and relate, but it is hard to see how biologist Stephen Jay Gould's "nonoverlapping magisteria"[16]—the view that science and religion do not conflict but have authority over domains that do not intersect—is tenable. One way of envisioning how they interact is the Dialogue view, described in the introduction as science and religion being different frequencies of light, each illuminating the world differently to increase the vision of both. Another framework is the Integration view, which sees science and religion as inextricably linked, where religion is the blueprint for what exists in the world. The compatibility of science and religion argued for and exemplified in this book is an appeal to ecumenism between disciplines and religious convictions for the advancement of science to the benefit of the human race. And this ecumenism is found even between the authors, one holding the Dialogue view, and the other the Integration view.

RICHARD C. POWELL

Like several of our "saints of science," I personally believe that science and religion are in dialogue as two parts of the same story. Science and religion properly understood agree. However, after 40 years of working with science colleagues who were mainly atheist or agnostic, I also appreciate that scientists do not always agree. Conflict is not between science and religion, but between people working in science and religion. It seems clear to me that the only way to overcome our differences is through open, honest discussion. Dialogue not only describes the relationship of science and religion, but the means of resolution between colleagues with whom there is disagreement.

DOUG POWELL

I am not a scientist, but a theologian and Christian apologist. I write on science only as a reporter with a philosophical eye.

Whatever the evidence is for evolution, I am proof that a talent for physics is not passed down genetically. I am theologically conservative with a high view of scripture. Yet, I do not think the Bible was written with scientific intent. I believe God revealed himself in scripture and in creation to form a testimony with a single, perspicuous voice.

The Hubble Ultra Deep Field (by NASA, ESA, H. Teplitz and M. Rafelski (IPAC-Caltech), A. Koekemoer (STScI), R. Windhorst (Arizona State University), and Z. Levay (STScI))

Look up. Look all around. Look inside yourself. Look into the past. Is everything you see and experience the product of random, undirected physical processes that were bound to occur given enough time? Or do you sense something beyond the physical, outside the reaches of what can be described by math or tested for in a lab? The same panoply of stars that can make us feel insignificant simultaneously declares the irresistible, shouted whisper of divine presence, immensely majestic and beautiful. Through telescope and microscope, the voice is the same: there is more to this world than can be quantified, measured, and observed. Science and religion properly understood increase our vision as we peer intently through the mirror darkly.

In benediction, the words of Dutch Prime Minister and theologian Abraham Kuyper (1837–1920) reveal the rhyme that unites science and religion:

> *"There is not one part of our world of thought that can be hermetically separated from the other parts; and there is not a square inch in the whole domain of our human life of which Christ, Who is Sovereign of all, does not cry: 'Mine!'"*[17]

References and Notes

1. Ecklund, E.H., et. al., "Religion among Scientists in International Context: A New Study of Scientists in Eight Regions," *Socius*, vol. 2:1–9. DOI: 10.1177/2378023116664353.
2. Collins, F., "In Praise of Harmony," a speech given in acceptance of the Templeton Prize, September 25, 2020.
3. Ibid.
4. Darwin, C., *Life and Letters of Charles Darwin, The*, Darwin, Francis, ed. (London: D. Appleton and Company, 1898), 41, 277–278.
5. Pascal, B., *Pensées*, 430 (New York: Dutton, 1958), 118.
6. Van Til, C., *The Defense of the Faith* (Philipsburg: Presbyterian and Reformed, 1979), 101.
7. Jastrow, R., *God and the Astronomers* (New York: Norton, 1978), 116.
8. Cf. Tertullian, Apology XXXIX, "But it is mainly the deeds of a love so noble that lead many to put a brand upon us. See, they say, how they love one another." Pliny, *The Letters of Pliny the Consul*, trans. William Melmoth (Dublin, Thomas Ewing, 1765), Book X, letter 97, 239–242.

9. Lieberman, J., *Faith's Answers to America's Political Crisis* (New York: Post Hill Press, 2024), XI.

10. Lewis, C.S., *The Abolition of Man* (New York: Macmillan, 1947, 1955), 95–121.

11. QuotedinCurtis,Adam,"AMileorTwooffYarmouth,"BBCblog,02.24.12. https://www.bbc.co.uk/webarchive/https%3A%2F%2Fwww.bbc.co.uk%2Fblogs%2Fadamcurtis%2Fentries%2F512cde83-3afb-3048-9ece-dba774b10f89 (accessed 01.16.24).

12. Ecklund, E.H., et. al., "Religion among Scientists in International Context: A New Study of Scientists in Eight Regions," *Socius*, vol. 2:1–9. DOI: 10.1177/2378023116664353.

13. Dawkins, R., *The Blind Watchmaker* (New York: Norton, 1996), 1.

14. Dembski, W., *Intelligent Design* (Downer's Grove, IL: InterVarsity Press, 1999), 127–134.

15. Cf. Tour, J.M., "Animadversions of a Synthetic Chemist," *Inference* 2, No. 2 (May 2016), https://inference-review.com/article/animadversions-of-a-synthetic-chemist (accessed 03.28.2025).

16. Gould, Stephen Jay, "Nonoverlapping Magesteria," *Natural History*, vol. 106, no. 2 (March 1997), 16–22.

17. Kuyper, A., *Sphere Sovereignty* (West Linn: Monergism Books, 2004).

Index